Blender 3D Asset Creation for the Metaverse

Unlock endless possibilities with 3D object creation, including metaverse characters and avatar models

Vinicius Machado Venâncio

BIRMINGHAM—MUMBAI

Blender 3D Asset Creation for the Metaverse

Copyright © 2023 Packt Publishing

Group Product Manager: Rohit Rajkumar

Publishing Product Manager: Vaideeshwari Muralikrishnan

Senior Content Development Editor: Feza Shaikh

Technical Editor: Saurabh Kadave and Simran Udasi

Copy Editor: Safis Editing

Project Coordinator: Manthan Patel and Aishwarya Mohan

Proofreader: Safis Editing

Indexer: Tejal Drauwale Soni

Production Designer: Vijay Kamble

Marketing Coordinators: Anamika Singh, Namita Velgekar, and Nivedita Pandey

First published: July 2023

Production reference: 1220623

Published by Packt Publishing Ltd.

Livery Place

35 Livery Street

Birmingham

B3 2PB, UK.

ISBN 978-1-80181-432-4

www.packtpub.com

To my mother, Claudia, who gave me unconditional love and support throughout my life and my journey as an artist. I wouldn't have made it this far without her.

– Vinicius Machado Venâncio

Contributors

About the author

Vinicius Machado Venâncio is a 3D designer with experience in asset creation in Blender for multiple purposes, ranging from animation to still renders to game-ready assets. As a freelancer, he's worked on a variety of projects for multiple companies and individuals, with various different styles, from realistic to stylized.

I want to thank the people who supported me throughout my journey, especially my mother, Claudia.

About the reviewer

Devendra R. Ghadge is a skilled WebGL developer based in India, with a keen interest in 3D modeling and graphics development. With significant industry experience in developing 3D models using Blender, he possesses sound knowledge of OpenGL and OpenGL-ES. Devendra gained all of his knowledge about 3D graphics from AstroMediComp, a renowned "Gurukul" that instills in its students a deep respect for technology and encourages a keen interest in learning. His education at AstroMediComp provided him with a solid foundation in 3D graphics. As a first-time book reviewer, Devendra brings his passion for technology and experience in 3D graphics to provide a unique perspective on the subject matter.

Table of Contents

7

Making the Base Mesh for a Humanoid Character   221

8

Refining the Base Meshes   257

9

Optimizing the Base Meshes   305

10

Rigging the Base Meshes   383

11

Further Development as a 3D Artist   447

Preface

Blender is a free and open source 3D modeling software that has been continuously growing in recent years, is extremely powerful, and is able to handle a lot of the tasks that most major 3D software out there also do. This book is a general overview of the tools used in asset creation available by default in Blender, as well as useful techniques to speed up the creation process and make the created asset as high quality as possible for a beginner.

Who this book is for

If you are a beginner 3D asset designer and/or character modeler and want to expand your skill set to asset creation for personal projects, games, and the metaverse, this book is for you. We'll cover in depth the process of creating such assets. Intermediate-level modelers could benefit from this as well. Basic texturing and UV unwrapping knowledge can help you get the most out of this book.

What this book covers

Chapter 1, *Modeling the 3D Asset*, guides you through the elements of the default interface and what they do. It then covers the basic tools and techniques used for modeling by going through the process of modeling a pair of headphones from scratch.

Chapter 2, *Optimizing Your Asset for Better Rendering Performance*, covers different ways of optimizing the asset made in the previous chapter if excessive geometry is present.

Chapter 3, *UV Unwrapping Your 3D Asset*, covers how to properly unwrap the optimized 3D model to prepare it for texturing, by first looking at primitive shapes and applying similar logic to unwrap the headphones. We will also cover how to manipulate the UV map to better occupy the 1:1 UV space.

Chapter 4, *Texturing Your 3D Asset Using PBR and Procedural Textures*, covers two distinct methods of applying textures to a 3D model: procedural, by using math to generate the necessary texture maps, and PBR, using pre-made texture maps.

Chapter 5, *Texture Painting and Using Real-Life Images as Textures*, goes over two more methods of texturing an asset: texture painting, which consists of painting the textures manually and offers absolute control over how the textures look, and using real-life images, in which we use and manipulate one real-life image to use as a texture and to generate the necessary maps.

Chapter 6, *Introduction to Blender's Sculpting Tools*, covers in depth the main tools Blender has available by default used for sculpting, such as the different brushes and how they behave when applied and their settings, to control that behavior.

Chapter 7, Making the Base Mesh for a Humanoid Character, covers the most important aspects of the human anatomy in order to make a believable character, going in depth into the different muscles, how they work, and how they affect the shape of their surroundings in the human body, as well as the general body proportions and the main differences between male and female bodies, which we'll use to start blocking out male and female base meshes.

Chapter 8, Refining the Base Meshes, goes over how to refine the base meshes we blocked out previously, adding detail and making heavy use of the anatomy studied in the previous chapter to add the necessary details and to make sure we have a decent, believable result.

Chapter 9, Optimizing the Base Meshes, covers the process of optimizing the geometry of the sculpted base meshes using a method known as retopology, where we reconstruct the body from scratch with a much cleaner topology, using the sculpt as a base. We'll also be covering how to recover the detail lost during the retopology process, followed by how to properly unwrap the base meshes.

Chapter 10, Rigging the Base Meshes, goes over the different types of movable joints in the human body, explaining their movements, limitations, and respective locations in the human body, to then add a rig to our optimized base mesh to move its different parts and limbs according to the limitations explained. We'll then cover how to add IK to different parts of the rig in order to pose it easily and how to generate a full-body rig using the built-in **Rigify** add-on.

Chapter 11, Further Development as a 3D Artist, suggests useful sources of knowledge in order to build up and improve the skills learned throughout this book.

To get the most out of this book

This book uses Blender 3.3 to make every asset. However, most techniques taught should work in newer versions.

Software/hardware covered in the book	Operating system requirements
Blender 3.3	Windows, macOS, or Linux

If you haven't installed Blender yet, you can do so at `blender.org`.

Download the color images

We also provide a PDF file that has color images of the screenshots and diagrams used in this book. You can download it here: `https://packt.link/8fJT4`.

Conventions used

There are a number of text conventions used throughout this book.

`Code in text`: Indicates code words in text, database table names, folder names, filenames, file extensions, pathnames, dummy URLs, user input, and Twitter handles.

Bold: Indicates a new term, an important word, or words that you see onscreen. For instance, words in menus or dialog boxes appear in **bold**. Here is an example: "Enable the **Rigify** add-on by checking the box to the left of the add-on's name."

> **Tips or important notes**
> Appear like this.

Get in touch

Feedback from our readers is always welcome.

General feedback: If you have questions about any aspect of this book, email us at `customercare@packtpub.com` and mention the book title in the subject of your message.

Errata: Although we have taken every care to ensure the accuracy of our content, mistakes do happen. If you have found a mistake in this book, we would be grateful if you would report this to us. Please visit `www.packtpub.com/support/errata` and fill in the form.

Piracy: If you come across any illegal copies of our works in any form on the internet, we would be grateful if you would provide us with the location address or website name. Please contact us at `copyright@packtpub.com` with a link to the material.

If you are interested in becoming an author: If there is a topic that you have expertise in and you are interested in either writing or contributing to a book, please visit `authors.packtpub.com`.

Share Your Thoughts

Once you've read *Blender 3D Asset Creation for the Metaverse*, we'd love to hear your thoughts! Scan the QR code below to go straight to the Amazon review page for this book and share your feedback.

https://packt.link/r/1801814325

Your review is important to us and the tech community and will help us make sure we're delivering excellent quality content.

Download a free PDF copy of this book

Thanks for purchasing this book!

Do you like to read on the go but are unable to carry your print books everywhere? Is your eBook purchase not compatible with the device of your choice?

Don't worry, now with every Packt book you get a DRM-free PDF version of that book at no cost.

Read anywhere, any place, on any device. Search, copy, and paste code from your favorite technical books directly into your application.

The perks don't stop there, you can get exclusive access to discounts, newsletters, and great free content in your inbox daily

Follow these simple steps to get the benefits:

1. Scan the QR code or visit the link below

https://packt.link/free-ebook/9781801814324

2. Submit your proof of purchase
3. That's it! We'll send your free PDF and other benefits to your email directly

Part 1:
Inorganic Asset Modeling

In this part, we'll go over the process used by most people when creating inorganic assets, also known as poly modeling. We'll cover each part in depth and make an asset using the techniques showcased from scratch that will be suitable for most uses, such as still renders, animations, and games.

This part has the following chapters:

- *Chapter 1, Modeling the 3D Asset*
- *Chapter 2, Optimizing Your Asset for Better Rendering Performance*
- *Chapter 3, UV Unwrapping Your 3D Asset*
- *Chapter 4, Texturing Your 3D Asset Using PBR and Procedural Textures*
- *Chapter 5, Texture Painting and Using Real-Life Images as Textures*

1
Modeling the 3D Asset

Blender is an extremely powerful tool for creating 3D models for a variety of purposes and has increased in popularity over the last few years mainly because it is a free and open source program, which makes it accessible to basically anyone with a computer. It has been designed so that every step of the 3D pipeline can be carried out within the program itself. That's why many people are starting to create assets for the metaverse using Blender rather than paid alternatives.

In this chapter, we'll learn how to start creating your own original assets, as well as learning how to navigate through Blender's interface and use its main modeling tools, using the modifier system to speed up the process.

In this chapter, we're going to cover the following topics:

- How to look for good image references
- How to break down an object into simple shapes
- How to create a base shape to build upon
- Deciding on the level of detail necessary
- How to gradually add detail to the base shape

Gathering references

Before starting any work, it is crucial that you have at least an idea of how the final product will look, as this will guide every other step of the process. Aspects such as the asset's style and materials are useful to keep in mind during the whole process. That's where image references shine, as they clearly show us how much detail to actually add to our models and what the general shape we should aim for is, among other aspects, which will make our asset look believable and fitting.

What are good and bad references?

When looking for image references, we come across thousands of images with different properties such as resolution, brightness, and lighting, and in order to guide ourselves correctly and effectively, we should be able to distinguish between what is a useful reference and what's not.

A good reference image should clearly show the shape, color, and material of an object in order to guide us correctly.

What to look for and avoid while looking for references

Generally, we'd want to avoid images with low resolution and bad lighting, which can flatten the shapes and "erase" detail. Photos taken with a light source located near the camera, for example, tend to flatten the shapes a lot, so it's a good idea to avoid pictures taken with flash or amatewur lighting. Pictures taken in a studio, on the other hand, tend to have ideal lighting, showcasing most of the details of the object.

Bad reference   Good reference

Figure 1.1 – Examples of bad and good references (images from https://www.pexels.com/)

From the preceding figure, we can notice several differences between the references: the bad reference doesn't show the general shape of the object clearly, as it was taken from a bad angle, while the good reference makes it very easy to see the silhouette as well as the sides of the headphones.

The lighting in the bad image makes it even harder to distinguish the shapes that make up the object, and the harsher, darker shadows "blend" the sides of the headphones with their own shadow, which doesn't help when defining the thickness of the headband. The lighting also prevents us from seeing the material properties, such as the surface bumpiness and roughness, clearly, especially in the cushions around the ear, as well as the plastic that makes the outer part of the headphones. The good reference, though, shows us the complete opposite: with clearer, brighter lighting and fewer postprocessing effects, it's now possible to notice most of the details on the surface, such as how shiny each part is; the headband and cushion materials are now clearly noticeable as they appear to be made out of a leathery material, along with some less shiny plastic parts.

Another aspect to keep an eye on is the exposure of the image as this can also make it harder to see details, so avoid under- or over-exposed images (see *Figure 1.2*):

Figure 1.2 – Examples of good and bad image exposures (image from https://www.pexels.com/)

It's worth noting that we should not work with one reference only, as this "limits" our view, so it's very important to find a set of reference images from a variety of angles and different versions of the same object (unless, of course, you're making a replica of a specific version), including of 3D models in your desired style, along with real-life pictures. PureRef is a good, free external tool, useful to organize the chosen reference images in a more concise way, though this is optional. Now, with a proper reference, we should be a nice guide to refer to throughout the entire process, which starts with blocking out the general shape of the asset.

Before we hop into making the actual headphones, though, let's have a quick look at the viewport so that we can familiarize ourselves with the environment we will be working in further down the road.

A quick tour of Blender

The first time you open Blender, you might feel overwhelmed just by looking at the interface:

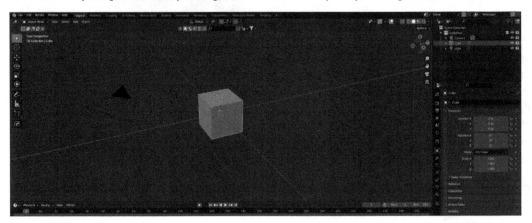

Figure 1.3 – Blender's default interface

Let's start with the bigger windows. The biggest one is the 3D Viewport, where you add objects, edit your models, and/or set up your scene overall.

In the upper right, we have the Outliner, which contains all objects in the scene and can help us locate different objects more easily.

Underneath it, we see the Properties Editor, where we can tweak several attributes and parameters of the currently selected mesh. It has several icons on its left side that illustrate what each tab inside of this editor affects.

Now, we can look at the Workspace menu at the very top of the Viewport:

Figure 1.4 – Workspace menu

There, we have many different workspaces, each one optimized with distinct interfaces for several different tasks, such as modeling, sculpting, and texturing.

Now, to the left of the viewport, we see some tools:

Figure 1.5 – Viewport tools

These tools, from top to bottom, are the following:

- **Select**: This tool allows for box selection and has more modes that can be toggled using the shortcut *W*.

- **Cursor**: This defines the location of the 3D cursor, which has a round red and white contour. The 3D cursor defines where the objects are added to the scene.

- **Move**: This allows us to move the object in one or more of the three axes.

- **Rotate**: This allows us to rotate the model in one or more of the three axes.

- **Scale**: This allows us to scale the object in one or more of the three axes.

- **Transform**: This is an all-in-one tool that provides the functionality of the three previous tools in one.

- **Annotation**: This allows us to make any type of annotation in our project.

- **Measure**: This measures the length and angle by clicking and dragging.

- **Add Cube**: This allows us to draw a cube or cuboid shape by clicking and dragging.

Note that most of these tools are easily accessible using shortcuts, and once you get used to using shortcuts, you will rarely find yourself using these tools.

Now, we can move to the right side of the Viewport:

Figure 1.6 – More tools

There we have a few other tools, again described here from top to bottom:

- **Options**: This is a menu with a few options for the current selection mode

- **Gizmo**: You can use this to figure out the view's orientation relative to the world coordinates and the scene, as well as rotating the view around by clicking and dragging

- **Zoom**: This zooms the view in or out by clicking and dragging

- **Move**: This pans the view

- **Toggle camera view**: This switches between the Camera and Orbit views

- **Perspective**: This switches between the Perspective and Orthographic views, which will be explained later in the chapter

While these may be tools for navigation in the viewport, there are easier ways of going about it: using shortcuts. The most common ones used for navigation are the following:

- *Middle mouse button*: Orbits the view around the scene.

- *Shift + middle mouse button*: Pans the view.

- *Scroll wheel*: Zooms in (scroll up) or out (scroll down). This can also be done using *Ctrl + middle mouse* button and dragging up or down.

For now, we will just mention the navigation shortcuts, but we will learn about more of them as we build out the headphones.

Blocking out the asset's general shape

After we have our references laid out and organized, it's time to start blocking out the general shape of the object so that we have a solid base to work on and add detail to later on. It's important to not focus on small details at this stage, as this is just our starting point. Think of this like a drawing: we start with a rough sketch and refine the forms and shapes gradually. This is basically the sketching part.

Breaking down the object into primitive shapes

In order for us to do our rough block out, we need to visualize the real-life object in simple shapes. Blender has a selection of basic shapes to start with that can be accessed through the **Add** menu by using the *Shift + A* shortcut and hovering the cursor over the **Mesh** option. Then, a panel should pop up with your options (see *Figures 1.7* and *1.8*).

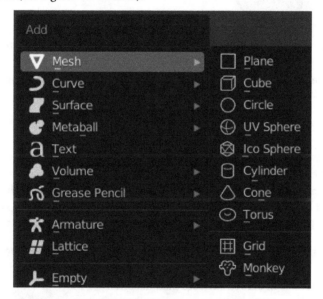

Figure 1.7 – Blender's main primitive shape options

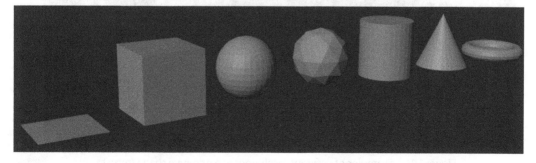

Figure 1.8 – All the main primitive shapes laid out

To start out, we need to use the available shapes that most resemble each part of our object by looking at the references we gathered before. So, let's separate our headphones into their individual parts so we can block them out properly.

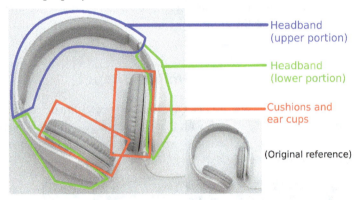

Figure 1.9 – Individual parts of a pair of headphones (image from: https://www.pexels.com/)

As we can see from the preceding figure, there are three main parts that make up the headphones, so now we have to decide on the **level of detail** (**LOD**) we want those parts to have.

How much detail do we need?

The amount of detail we will put into our model heavily depends on factors such as how far away the object is from the camera, the desired style, and how important it is in our metaverse game. That's why many projects (especially games) include several versions of the same model with varying amounts of detail, the so-called LODs – a higher-detail version for when there is a close-up of the model and a lower-detail version for when it's far away.

Let's say, for example, that we want a blocky and geometric look for our headphones; then, we would use lower-resolution primitive shapes rather than the ones that Blender provides by default. The resolution is more noticeable in curved surfaces than flat ones.

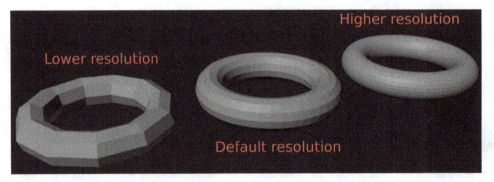

Figure 1.10 – Examples of different resolutions in a primitive shape (torus)

As we can see, the more resolution we add, the more detailed the surface looks, which can provide a more realistic look to our object.

It's very important to note, though, that the more resolution we opt to put in the asset, the harder it will be to render, so we need to find a balance between the resolution and the look we're going for.

For our example, we'll be going with a semi-realistic look, so we might need a bit more detail than the default settings Blender provides us with. Now we're ready to actually start modeling our headphones.

Building the headphones inside Blender

For this model, we're going to start with the cushions and earcups, which can be simplified into one part that resembles a cylinder pretty nicely.

Creating the lower part of the headphones

When you first open Blender, you'll see three default objects: a cube, a sun light, and a camera. We won't be using them, so we'll start by deleting all of them by pressing A to select all of them, followed by either *Delete* or *X*, then selecting **Delete** in the menu that pops up.

Now we'll proceed to add a cylinder. The default cylinder has 32 faces on its side, which is enough resolution for us considering that it is a relatively small object, so let's stick with that. After adding the cylinder using the previously mentioned shortcut for adding objects (Shift + A), this is what we see in the viewport:

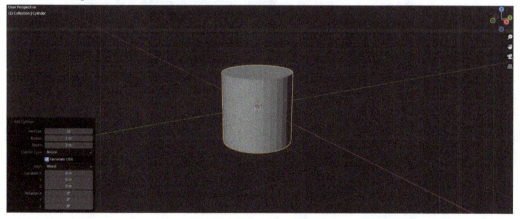

Figure 1.11 – The default cylinder

Immediately, we notice that this cylinder is way too thick and oriented the wrong way. So, we will start shaping it into our base model. First off, we need to see our object from other angles, which can be done by using the middle mouse button to rotate the view. Holding *Shift* while doing so pans the view around. Zooming in can be done by scrolling.

Secondly, we need to rotate the object to the side. For that, we use the *R* shortcut on the keyboard. Then, we could adjust the rotation by dragging our cursor around the object, but that's imprecise. To rotate it in a more precise way, we can limit the rotation to either the X, Y, or Z axis by typing the relevant letter after pressing *R*. To know which is the right axis, we can look for the little orientation gizmo in the top-right corner (*Figure 1.12*).

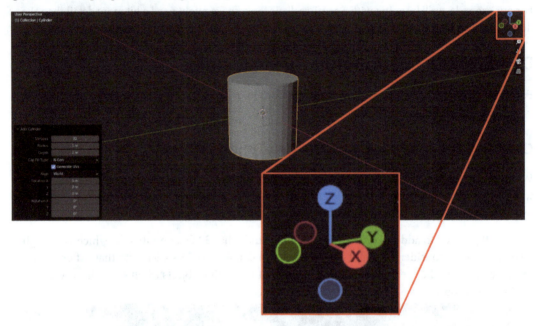

Figure 1.12 – Blender's orientation gizmo

Once you've constrained the rotation for the desired axis (in our case, it's the X axis), you can type the exact angle you want it to rotate, which is 90° in this case, and left-click to confirm your action (right-click cancels it). Then, you should have something like this:

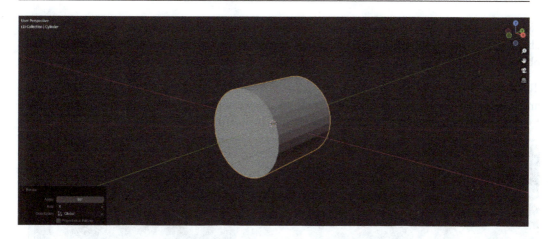

Figure 1.13 – Cylinder rotated 90° on the X axis

Now, we need to scale it down to the desired thickness by using the *S* shortcut on the keyboard, again constraining the axis so we don't scale it all at once, and dragging the cursor toward or away from our model to adjust the size (you can also type the scale using keyboard keys just like rotating, but scaling works in a multiplicative way rather than additive. We'll drag the cursor for this one as it doesn't require a lot of precision). In order to see the changes better, we can press the numbers *1*, *3*, or *7* on the numpad (or the ~ key, then set the view if you're on a Mac) to snap our view to the front, right, and top views, respectively (you can press *Ctrl + 1*, *3*, or *7* to go to the opposite view). This is what we have now:

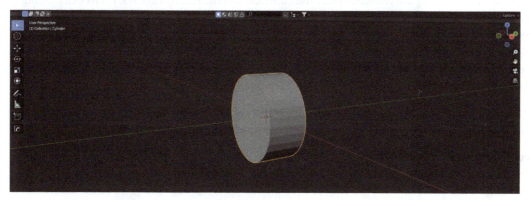

Figure 1.14 – Cylinder resized

It's important to note that after you make any changes to the scale, you should apply the new scale by pressing the *Ctrl + A* shortcut and selecting **Scale** from the menu that pops up. This will prevent problems when we make the cushions later on.

Now, it's time to introduce you to where more detailed shaping happens, as well as where you'll spend about 50% of your time: edit mode, which can be acessed by pressing Tab while with the desired object selected.

That's where we can edit the individual vertices, edges, and faces of our geometry, as well as adding and removing them when necessary. We can toggle between these modes using the buttons in the top-left corner or by pressing *1*, *2*, and *3* on a normal keyboard (*Figure 1.15*).

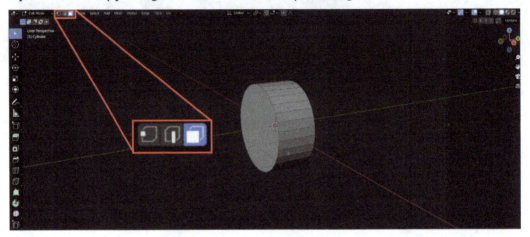

Figure 1.15 – Vertex, edge, and face selection modes

To start, we can see in the references that the cushions and the earcups are not perfect circles, but rather have a slightly oval shape. To do that, we select all faces using the *A* shortcut on the keyboard and scale the earcup on the correct axis (Z) until we are satisfied with the shape. If you're following along, you can scale it up or down according to your taste. It's worth noting that you can reshape this at any point but it will become harder to do the further along you are in the process. That's why we focus on the bigger shapes first and add the smaller details later.

After that, it's time to take care of faking the separation of the earcups and cushions, as we chose to keep them in one shape. To do that though, we need to add more geometry to our model.

For this, we're going to add what's called an "edge loop," which is a ring of edges that will go around a set of faces, dividing each one of them into two. To add an edge loop, we do the following:

1. Go into edit mode.

2. Press *Ctrl + R.*

3. Hover our cursor over the ring of faces we want to add our edge loop on.

4. Click with the *left mouse* button.

5. Drag our cursor to where we want our loop to be (where the cushions end and earcups begin) and click again.

After that, you should have something that looks similar to this:

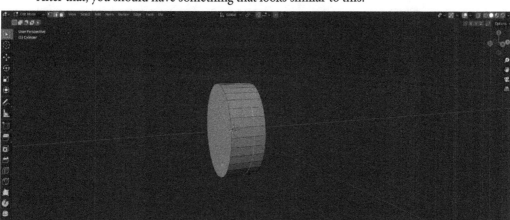

Figure 1.16 – Edge loop added

Now it's time to shape the earcups and the cushions, respectively. To do that, we do the following:

1. Select the round face with the shortest faces connected (if you want to tweak more than one face at once, hold *Shift* while selecting to select multiple faces).

2. Scale it down to the desired size (if you want, you can readjust the location of that face by pressing *G* and then dragging your cursor. Restricting the axis is also possible).

3. Add another loop to the left of the initial loop (that's going to be the base of the cushion).

   It's worth remembering that *Ctrl* + *Z* undoes the last action if you are unsatisfied with a change you just made. After that, you should have something that looks like this:

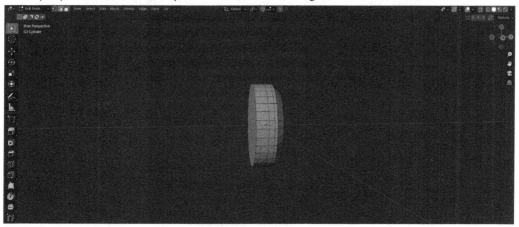

Figure 1.17 – Earcups adjusted

We need to pull the base of the cushions outward to shape it, which can stick out in some headphones. For that, we'll need to extrude it. Regular extrusion (the *E* shortcut) will pull all the faces in one direction, but what we want is to pull each face in the direction that it is pointing in, which is called the normal. To access the different types of extrusion, we need to select the face loop (shortcut: *Alt* + left-click on the loop using the face selection mode), go into the extrusion menu by pressing *Alt* + *E*, and from the menu that pops up, select **Extrude faces along normals**. Now you can drag your cursor and adjust the extrusion to your liking. Your model should look like this:

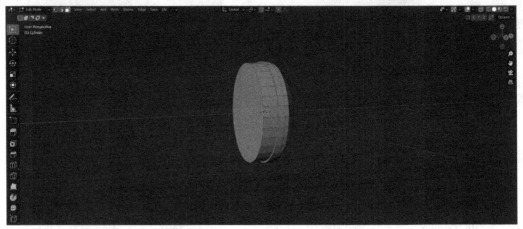

Figure 1.18 – The base of the cushions extruded

The next step will be to add the hole in the middle of the cushions where our ears go. To make that hole, we'll inset the front face using the *I* shortcut and drag the cursor, and a new ring of faces should appear with a new round face in the middle. Adjust it to a size slightly smaller than the one of the actual hole, as in the next step when we round the edges of the cushion, it will look bigger. After that, you can extrude the round face inward using regular extrusion, as we want it to go in just one direction. This is what you should have now:

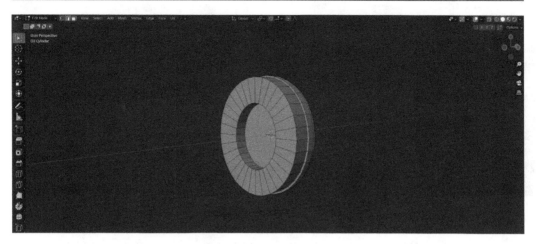

Figure 1.19 – Carved hole in the cushions

For the last step of this part, we'll round the corners of the cushion in order to make it look more like a soft surface. We'll start by going into edge select mode using the shortcut *2* and selecting the edge loops that we want to be rounded (the inner and outer edges of the cushion in this case):

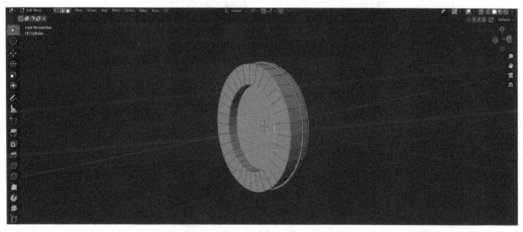

Figure 1.20 – Edges that will be rounded selected

Now, we'll press the *Ctrl* + *B* shortcut to add a bevel and drag the cursor to adjust the size. Initially, your bevel will have only one face, which isn't enough to make it look round. While adjusting the size of your bevel, you can scroll up or down with the mouse wheel to increase or decrease the resolution of the bevel to your liking, and if you want it to be round all the way, you can press *C* to clamp the bevel while adjusting it. This will prevent the vertices from overlapping with each other by limiting the bevel amount we can apply, which causes several issues. You should end up with this:

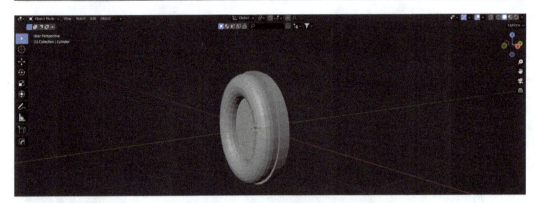

Figure 1.21 – Cushions rounded

At this point, it's a good idea to save your project using *Ctrl + S*, to guarantee that you don't lose your progress and your sanity if Blender crashes in the middle of something. Saving regularly is good practice.

We're finished with blocking out the earcups and cushions. Now it's time to move on to the bottom part of the headband (see *Figure 1.22*).

Figure 1.22 – Bottom part of the headband (image from: https://www.pexels.com/)

To start off, we won't add another primitive shape, as we will use the earcup's contour to shape that part. So, while still in edit mode, start by selecting about half of the upper ring of faces that go down the earcup's side and duplicating it using the *Shift + D* shortcut. Then, we need to drag those faces along their normals to around the thickness we want that bottom part to be, and after that, separate the earcups from this headband part. To do that, we do the following:

1.   Select all the separate faces.

2.   Press the *Alt + S* shortcut.

3. Drag your cursor up or down to adjust the thickness.

4. Left-click to confirm.

5. Press the *P* shortcut to separate the faces and select **By Selection** from the menu that pops up.

Now you should have two separate objects in your scene when you go out of edit mode and back to object mode in your viewport (1), as well as in your outliner (2), as seen in the following screenshot:

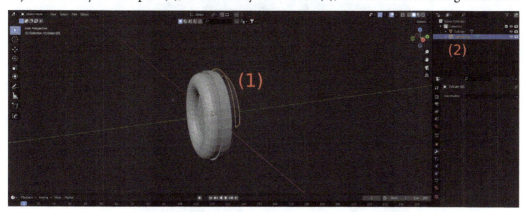

Figure 1.23 – Earcups and headband separated

As you can see, this part of the headband is infinitely thin and leaves a gap between the parts, so we need to add thickness to it. We could extrude the faces along the normals again, but instead let's not; we're going to use Modifiers. Modifiers are a way to preview changes in our model but alter them later on if we feel like it. The modifiers tab can be found by selecting the blue wrench icon to the right of the viewport:

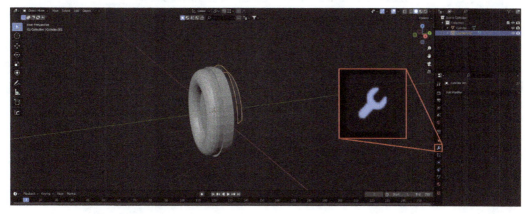

Figure 1.24 – Solidify modifier settings

In that tab, when we select **Add Modifier**, we have access to a bunch of modifiers that we can use. Most of the names are self-explanatory and have an icon to their side illustrating what each one does. For now, we'll go only with the Solidify modifier, which can be found under the **Generate** column, which only has modifiers that alter the geometry. Upon selecting **Solidify**, you will be able to see the modifier settings appear in the modifiers tab.

Figure 1.25 – Solidify modifier settings

The main settings are **Thickness**, **Even Thickness**, **Fill**, and **Only Rim**, but we're only going to use **Thickness** and **Even Thickness** for our headphones. Click and drag the **Thickness** slider left or right until it touches the earcup's sides and check the **Even Thickness** option.

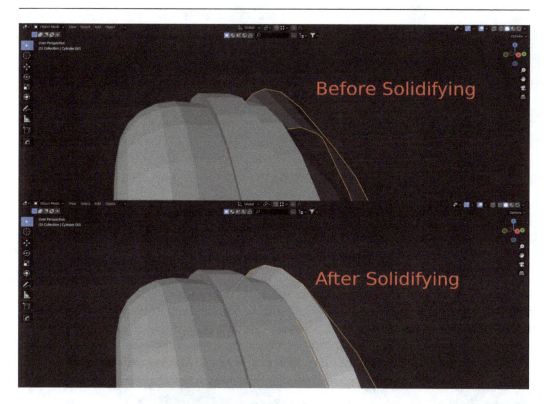

Figure 1.26 – Before and after solidifying

The next step is to extrude the part that connects the bottom part to the top part of the headband, and thanks to our solidify modifier, the edges we extrude and/or tweak are going to maintain their thickness.

Start by selecting the edges you want to turn into the connecting part (2-6 edges ideally, if you're following along with similar scale). Remember to keep the selection symmetric and extrude the edges up to the desired height. You can also tweak the location of those faces to your liking, avoiding overlaps of course. It should look something like this:

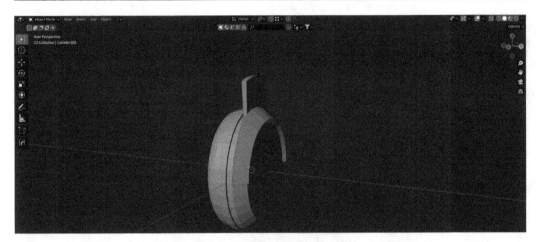

Figure 1.27 – Base of the extrusion

That's the base of our extrusion, but most of the references show a slight curve in the extrusion. To do that, we're going to flatten the top area by selecting the edges and scaling them in the Z axis by 0 (this multiplies their relative scale by 0, effectively aligning them perfectly), then adding a single edge loop with *Ctrl + R* and then right-clicking to place it exactly in the center. Then, pull it back along the Y axis until you're satisfied with the curve. After that, bevel that edge loop to smooth the curve.

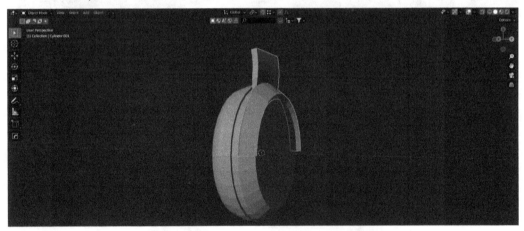

Figure 1.28 – Curve added to the connecting part

For extra control over the thickness, you can apply the solidify modifier by hovering your cursor over it in the modifiers tab and pressing *Ctrl + A*. This will apply the geometry generated by the modifier and let you manually control the thickness of each face. If you want to keep your changes perfectly symmetric, you can do the following:

1.   In edit mode, select the half you want to symmetrize your changes to.

2. Delete it using the *Delete* or *X* shortcut, then select either **Vertices**, **Edges**, or **Faces** from the delete menu (sometimes not all deleting modes will give out symmetric results, depending on the model, so choose what works).

3. Add a **Mirror** modifier set to the X axis (in this case) and make sure merge and clipping are enabled. This will make sure the object is just one piece and has no vertices in the exact same position (see *Figure 1.29*).

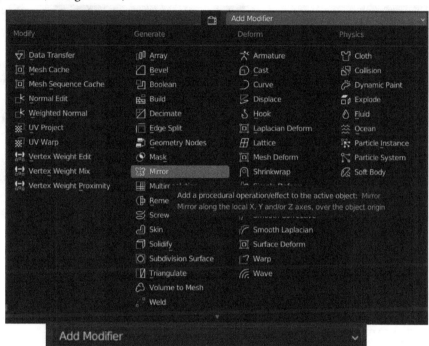

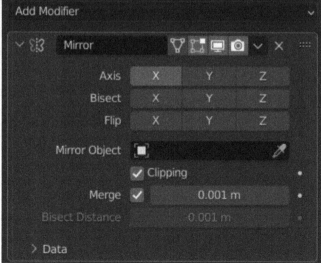

Figure 1.29 – Mirror modifier

If your mirror doesn't look right, you can try to change the mirror axis, and if that doesn't work, undo the actions until the deleted faces come back. It's important to remember that the mirroring is done based on the location of the object's origin, which is a tiny orange dot (see *Figure 1.30*). So, make sure that it is in the middle of the object (in the desired axis). Many things are tied to the origin of the object, such as rotation, simulations, and many modifiers, so it's important to keep it wherever we want it to be.

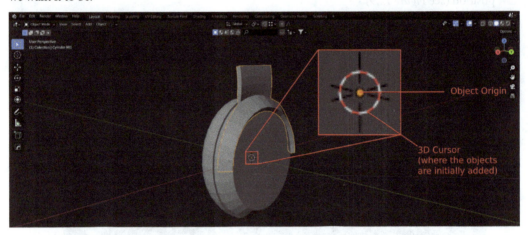

Figure 1.30 – Object origin

If you need to change where the origin is, you can do it by doing the following:

1. Select the object that you want to recalculate the origin of.
2. Right-click to open the object context menu.
3. Hover your cursor over **Set Origin**.
4. Select **Origin to geometry**. This will move the origin to the median point of the model based on the density of your faces, edges, and vertices.

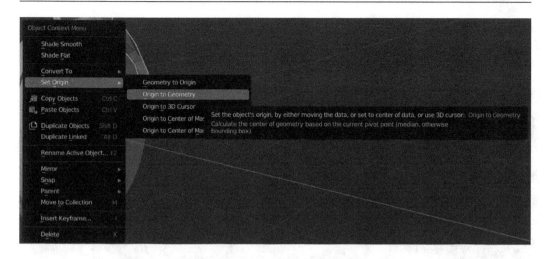

Figure 1.31 – Setting the origin to the median point

Now that your origin is correctly placed, everything that depends on it to work should work as expected. So, you can again delete half of the geometry in edit mode and add the mirror modifier (if it failed on the first try), and tweak the shape as you desire. If you're going to tweak the shape, you might want to merge or connect one vertex to another to create or get rid of a face or edge. To fill the space in between them with an edge, you can select two vertices and press *F*, but if there's a face already connecting the vertices , you'd want to cut that face (which you can't do with the fill operation). So, in this case, you'd connect the two selected vertices by pressing *J*. To merge them into one, though, select the two vertices you want to merge with *Shift + left-click* and use the shortcut M, then choose an option for the menu that pops up:

Figure 1.32 – The Merge menu

After you reshape your model to your liking, you can either apply the mirror modifier or keep it there if you plan to make symmetrical changes or add details to the geometry later on. This is what the final earcup block out should be similar to (remember: smaller details come after the block out stage):

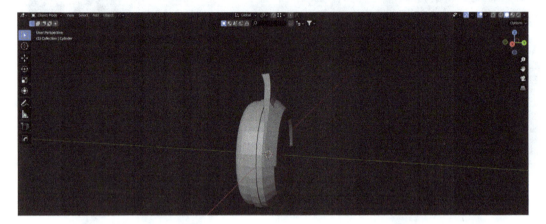

Figure 1.33 – Finished block out of the earcups, cushions, and bottom headband part

It's now time to move on to the last two pieces of our headphones: the upper headband and its cushion.

Creating the headband

We can start the same way as we did for the bottom part: by again duplicating the top faces of the lower part in edit mode, separating them by selection, and dragging slightly upward, though this time in the normal direction to maintain the alignment, as the faces are slightly tilted. To change the transform orientation, select the orientation menu in the top-middle of the Viewport, then select **Normal**:

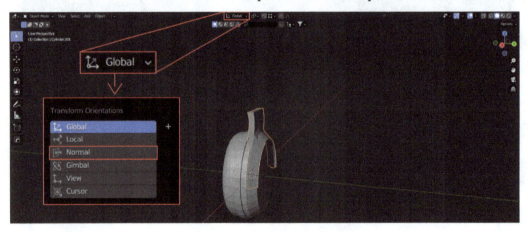

Figure 1.34 – Changing the transform orientation from Global to Normal

Now, when we drag the duplicated top faces on the Z axis, they should remain aligned. After this step, we have our base and you can go back to the **Global** orientation.

We could extrude it and rotate/drag each extrusion into place, but this takes too long and has a very high chance of not looking very circular, so let's let our computer do the hard work for us with the Spin command (which currently doesn't have a shortcut by default).

To use it, we have to go into edit mode and select the faces we want to extrude around a center point, and in the toolbar to the right (which also contains all other edit mode tools we talked about up to this point), select the Spin icon:

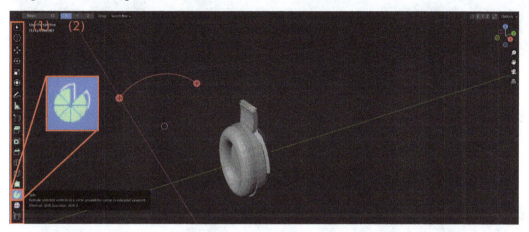

Figure 1.35 – The Spin tool

Before spinning the face, it's recommended that you apply the mirror modifier and turn the four remaining faces into one by selecting the middle edges and pressing *Ctrl + X* to dissolve the selection while still keeping the face there.

After that, when you select the remaining face, an arch will appear around the 3D cursor (which will be the center point of the extrusion), as well as the settings for the steps *(1)*, which is the number of faces the generated arc will have, and the axis of extrusion *(2)*.

Now, we have to position the 3D cursor in the right place using the *Ctrl + right-click* shortcut with our mouse to where we want it to be. We need to imagine the headband as a full circle and place the cursor roughly in the center of it. It's very easy to get it wrong on the first try and notice the incorrect placement only when you start spinning it, so you can always undo this operation with *Ctrl + Z* and fine-tune the position until it looks right. The right place should be around here:

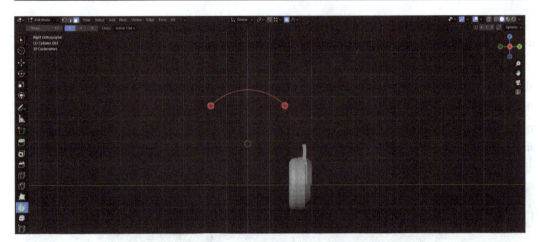

Figure 1.36 – Correct location for the 3D cursor

To begin our extrusion, we click on one of the + icons in the arch and drag it to the side we want to extrude it to. Holding *Ctrl* while doing so will snap the extrusion to 15° increments (we're aiming for a 180° or -180° extrusion in this case, depending on which side of the headphone you started with). It's important to note that the default step amount is a bit too low for us, so we're going to need to increase it until it's enough, which is around 48 in our case. This is what your headband should look like now:

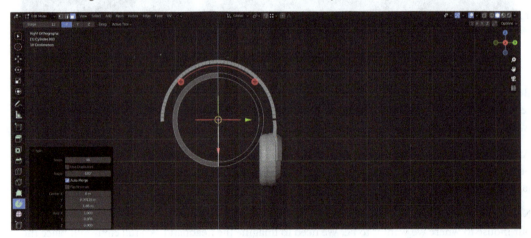

Figure 1.37 – Circular extrusion made

As you can see, most of the headband is created for us with this simple command. For extra control over the shape, you can play with the settings and sliders in the menu that pops up on the bottom left.

Looking at the references, we can see that the headband is actually not perfectly circular, but rather flattens a tiny bit at the top, so that's what we're going to implement now. We could move each face individually by hand, but this would take too long and has a high chance of not looking good at all (again!). So, let's let our computers do the hard work again with the proportional editing tool, which can be toggled either by pressing *O* or by clicking the icon on the top-middle of the Viewport:

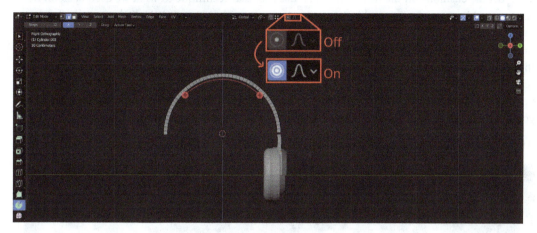

Figure 1.38 – Proportional editing

Now, every time we drag, rotate, or scale a face/edge/vertex, the geometry around it will be affected by a falloff. With proportional editing active, select the top-middle faces of the headband and drag them down in the Z axis. Once you press *G* to grab, a circle will appear around your cursor. This is your falloff size, and you can control it by scrolling up and down. Once you drag that middle face down, you'll see the geometry around it go down too. After that, you should have something like this:

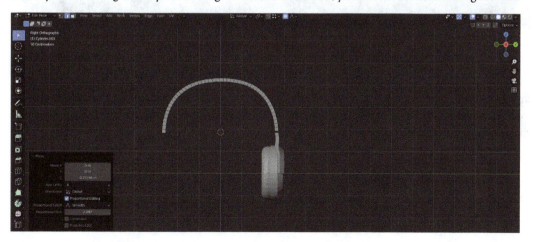

Figure 1.39 – Top of the headband flattened

If we wanted to add our mirror modifier back again to keep it symmetrical, we can actually delete three-quarters of the headband since headbands tend to be symmetrical on both axes (you can set more than one axis for just one mirror modifier; in this case, we'll use X and Y). That way, we only have to model a quarter of the object.

Tomodel the cushions on the headband, we'll use a similar technique to what we used back on the earcup's cushion to make it seem like two separate pieces: with proportional editing turned off, select the faces that you want to turn into the cushion, inset them using *I* (you can press *B* while insetting the faces to make the inset border take the mirror modifier into account if you decided to add it after the spin command), and extrude it along the normals using the *Alt + E* shortcut. If you want, you can scale the bottom faces down in the X axis to make it narrower than its base. You can also scale them up or down along the normals to make the cushion thicker or thinner after you extrude. After that, you can round the corners to your liking by using *Ctrl + B*, just like the earcup cushion.

After that, our blocking out is 98% done, and you should have something like this:

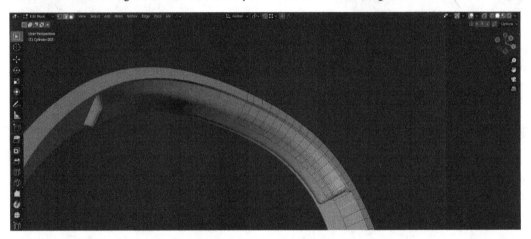

Figure 1.40 – Headband cushion finished

Alright, now the modeling for this part is done! It's still missing the other earcups and bottom headband though. To add them, we can add a mirror modifier to both of the objects and set the mirror object to the headband in the Y axis using the little eyedropper icon in the modifier settings (click on it and click on the object that it's supposed to mirror across):

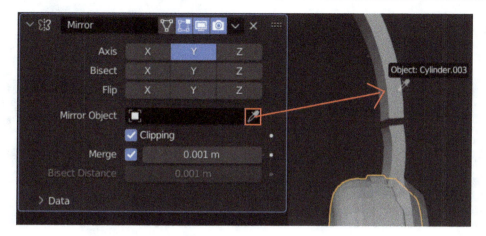

Figure 1.41 – Mirroring objects across the headband

OK, now we have both sides ready, so we can position those to our liking: some headphones have a slight tilt to those parts when not being worn by someone, so we can add that here by grabbing/rotating both objects at the same time. This is the result:

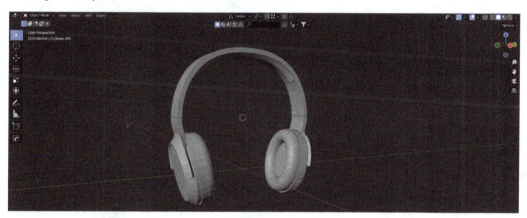

Figure 1.42 – Finished block out

One last thing before we move on to the details: notice how everything looks blocky, with the faces and edges clearly visible. This could be fixed by adding more geometry but then the model wouldn't be suitable for a game. Instead, we will tell Blender to fake the smoothness using smooth shading.

To apply smooth shading, we select the objects we want to add it to and right-click, then select **Shade smooth**. The problem with this is that it smoothes out everything, even the sharp edges. To fix that, we go to the **Object data properties** tab and under the **Normals** option, check **Auto smooth**. This will smooth the shading based on the angle between two edges, which can be tweaked (though 30°-60° tends to work well in most cases). After that, the final block out should look like this:

Figure 1.43 – Block out finished

It's worth remembering that it's possible to reshape, tweak, and add more shapes to the model, such as cat ears or a microphone. At this point, you know most of the main Blender commands and shortcuts and should be able to add objects by yourself.

Perfect, now it's time to add the much-appreciated details!

Refining the block out to add detail

With our block out now complete, we should start adding details to our model. This process, just like the one before, consists of focusing first on the bigger details, and then going on to smaller and smaller ones (if you need/want them). Those are called primary, secondary, and tertiary details. We already completed the primary phase, which mostly focuses on the silhouette and general structure.

Secondary details

This type of detail consists of medium-sized shapes that add complexity and interest to the bigger shape – things such as metal plates, tubes, and bigger carvings in walls. Although still relatively big, they play a big role in making a model look nice.

Though the idea of adding detail everywhere can seem very good, it's very easy to add too much detail, and even worse, adding too much detail to one part while leaving the others empty.

Here are some suggestions of secondary details to add to our headphones:

Figure 1.44 – Secondary detail added to the earcups

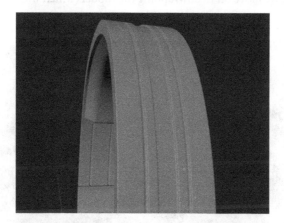

Figure 1.45 – Secondary detail added to the headband

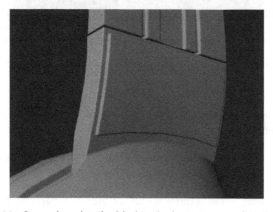

Figure 1.46 – Secondary detail added to the bottom part of the headband

As you can see, the secondary details added made our headphones look less "bland" and generic. Rounded edges were also added with a bevel modifier, but if you plan on doing the same, be very careful as it adds a significant amount of geometry, which can make the model too heavy for a game: what once was a headphone with only around 3.5k faces turned into nearly 11k faces just by adding the bevel modifier set to 3 faces.

It does look nicer though, so you need to find a balance between looks and the object's importance in the game. For example, if the model is going to be seen from far away, you probably don't need the rounded edges, but if it's going to be very close to the player, you can use more detail if your object needs to look realistic. We're going to keep the bevel modifier for looks but you might not want to add it (it won't make a big difference to the process).

We should avoid applying the bevel modifier as much as possible for now, as applying it will make our life way harder when we move on to unwrapping the model (which will be explained later on).

Here are the full headphones with secondary details added:

Figure 1.47 – Full headphones with secondary details

Tertiary details

This type of detail is composed of smaller shapes such as bolts, small wires, smaller indents, small holes… small shapes in general.

As with any type of detail, some sort of balance should also be struck with its usage throughout the model; note that not every part of the object will need this LOD.

As for our headphones, tertiary detail is more useful for the lower parts, so here is a suggestion for what could be added:

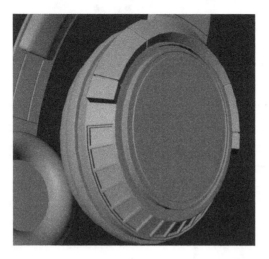

Figure 1.48 – Tertiary details added to the earcups and bottom part of the headband

This was done using the same techniques explained before, except for the grid-like structure on the earcups, which were done using individual face insetting, by pressing *I* twice when insetting more than one face. Then, we can extrude the faces along the normals as usual. After that, this is what the full model looks like:

Figure 1.49 – Final model after adding tertiary details

Congratulations, you finished the modeling part of your asset! Now, we might need to optimize it to make it lighter to render.

As a quick reminder for future projects, here's a small table with a few of the most useful and common modeling shortcuts learned by now:

Shortcut	Function
Tab	Switches between viewport modes
Shift + *A*	Adds an object
G	Grabs the selected objects/faces/edges/vertices
S	Scales the selected objects/faces/edges/vertices
R	Rotates the selected objects/faces/edges/vertices
I	Insets the selected faces
E	Extrudes the selected faces/edges/vertices
Shift + *D*	Duplicates the selected objects/faces/edges/vertices
Ctrl + B	Bevels the selected edges
Ctrl + A	Applies a parameter, selected in the menu

There are many more shortcuts, and some of them will be covered and used throughout the next chapters, but these are some of the most used ones.

Summary

In this chapter, we covered how to model an inorganic asset from scratch, as well as learning about most of the main inorganic modeling tools available by default in Blender and where to add the different LODs in an additive way. This is the first part of creating any 3D model and sets the LOD of the next steps, such as texturing.

In the next chapter, we will cover optimizing our model to make it lighter in terms of geometry so that we can render it more easily.

2

Optimizing Your Asset for Better Rendering Performance

Now that you have successfully applied 3D modeling to your asset, it's time to make it lighter in terms of geometry to render it more easily inside your project, game, or metaverse world. To do that, we'll need to see whether our model has an excessive number of faces, and if it does, we'll learn how to deal with such cases.

This process is optional, though, as most inorganic models tend to need much less or even no optimizing (depending on the purpose); hence, this process is only applicable to models containing more geometry than necessary for the object's purpose, size, and importance in-game.

In this chapter, we'll cover the following topics:

- How to detect excessive geometry
- Different ways to deal with excessive geometry

Deciding whether an asset has excessive geometry

Before optimizing our 3D model, we first need to check and decide whether it really needs it. Generally, inorganic non-deformable assets will need much less optimization work if the asset's purpose is taken into account at the beginning. Our headphones, for example, wouldn't need much optimizing for still renders or animation, but we'll have to optimize them further to make them work in games; however, for demonstration purposes, excess geometry will be added.

We'll first optimize them to the point where we concluded in the last chapter; then, we'll cover the additional optimization we'd need to do for games.

To detect excess geometry, we should look for edges or edge loops that don't contribute to the surface detail or curvature and dissolve them with the previously mentioned *Ctrl + X* shortcut.

Reducing excessive geometry

There are different ways to optimize a 3D model to achieve better rendering performance. In non-organic modeling, two main methods of dealing with such issues are dissolving edges and un-subdividing the mesh. The first method is done manually, which can take longer, but you have total control over the result; therefore, this approach can be applied to almost any inorganic asset. The second method is automatic and takes seconds, but you have no control over the result, and therefore, the end result may look good or bad depending on your model and its geometry. We'll apply the first technique to the cushions of the headphones.

We recommend making a copy of the high-poly objects first, so we can, later on, take all of the detail from the higher-resolution model into the optimized one, which is particularly used in in-game models. So you can leave all the modifiers you may have applied to make it higher quality (for example, **Bevel** and **Subdivision surface**), as this detail will be transferred later when baking the textures for games.

Dissolving edges manually

Let's say that you decided that you needed more resolution for your cushions when we beveled them, but the number you picked was a little too high now that you think about it since the object will be relatively small. We will have to dissolve some of those edge loops to make it lighter in geometry. The following shows what we have now in Edit Mode:

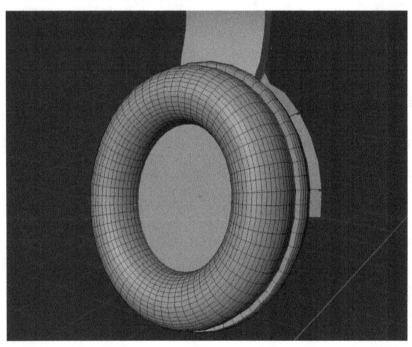

Figure 2.1 – Cushions with excessive geometry

As you can see, the cushions have much more geometry thaan necessary, with around 3,600 faces, and the curves won't be affected nor look blockier if we dissolve some of those edge loops, which we can do to several of them. The process is relatively simple, though a bit time-consuming.

It's important to know that the geometry has to be dissolved uniformly to maintain consistency throughout the curves so that one part doesn't look like it has a higher resolution than the others.

A good technique is to select every other edge loop and see how it looks after dissolving them. Of course, if we select a wrong loop or the result doesn't look as we want, we can always use the good old *Ctrl + Z* shortcut.

Let's start, then. To select multiple loops at a time, we have to hold *Alt + Shift* while we select our loops; clicking again on a selected loop deselects it. This is what we should see after we select the loops we want:

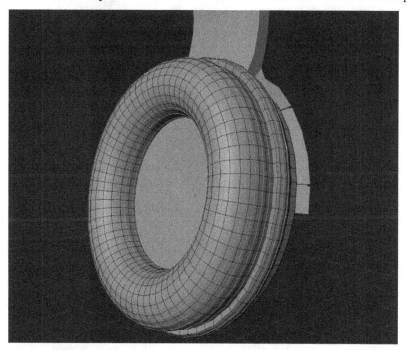

Figure 2.2 – Every other edge selected around the cushions

Now that we have selected all the edges we initially want to dissolve, it's time to dissolve them using *Ctrl + X*; we should see a result similar to the following once this is done:

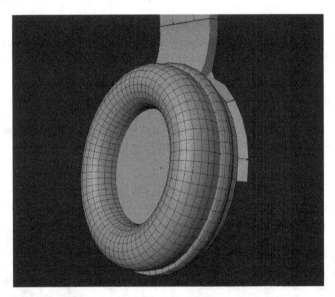

Figure 2.3 – Edge loops deleted around the cushions

We just dissolved around a thousand faces (in this case), which is not a lot if you think about it, But if we extend the issue to a big scene with lots of objects, for example, this can significantly impact performance.

Great, but we can go further. Notice that quite a few loops going in the other direction can also be deleted. We can delete them using the same technique. This is what your selection should look like now:

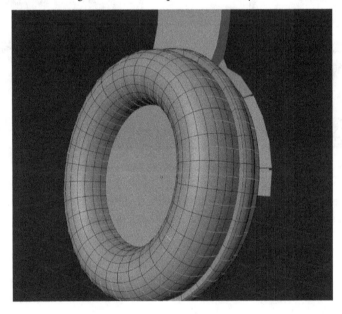

Figure 2.4 – Every other loop going the other way selected

Now we can dissolve them with *Ctrl + X*, just like we did before:

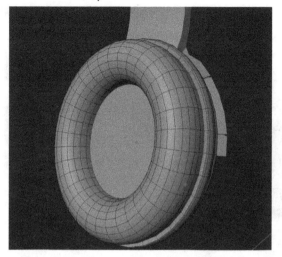

Figure 2.5 – Excessive loops deleted

By dissolving those last edge loops, we effectively reduced the number of faces by around 1,200, which leaves us with around 2,200 faces. We can go further, though.

Lines that are supposed to be straight can be composed of just one face in length, and upon further inspection, we can see that two edge loops have little to no effect on the overall curvature, making it possible to dissolve them and maintain the surface curvature virtually the same, or more specifically, the last inner and outer loops:

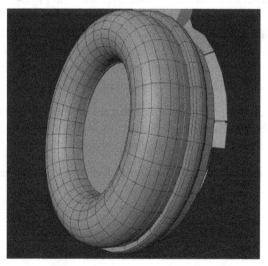

Figure 2.6 – The last inner and outer loops

When we dissolve those loops, the face count goes down by around 70, which is not a lot, but it's better than nothing, and it will make life a bit easier for us when we prepare our model for texturing later on.

The following screenshot shows the before and after optimization comparison for the earcups:

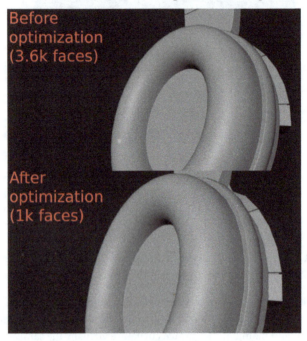

Figure 2.7 – Before and after optimization face counts and visual difference

As you can see, we effectively reduced the face count by two-thirds of its original value with little to no visual difference. However, we can see that the outer part of the cushions looks more blocky, with the edges more visible than before optimization. It's worth noting that any number shown doesn't consider the bevel modifier, which we retained in our model, because it makes it easier to see the edges.

This can sound like a negative factor, but remember that the object is small; therefore, it will mostly be seen from a distance, which means that less geometry is necessary to convey the same roundness. Plus, we can add, dissolve, and tweak as many edges as needed.

With the earcups and cushions optimized, it's time to talk about the second method of optimizing a 3D asset.

Decimation

Let's say that, while playing around with modifiers, you stumbled upon one of them called **Subdivision surface** and decided to give it a try, and seeing how smooth and high resolution it made the whole model look, you decided to keep it in the model.

The problem is, though, that while this does make the model look a lot less blocky and more realistic, it adds a lot of unnecessary faces, as it splits each face into four and averages the position of the vertices to make it smoother, which can make your geometry look very, very dense very quickly as you increase the subdivisions:

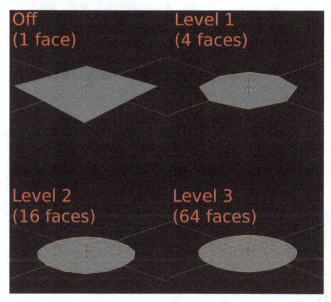

Figure 2.8 – Different levels of subdivisions

Generally, the subdivision surface technique is used in still renders, animation, and characters and is not so commonly used for inorganic models that will have to be run at 60 frames per second or more.

Therefore, if you decide to use it and have already applied it, this optimization method can give you some nice-looking results. For example, let's say we applied two levels of subdivision to our original headband, combined with the bevel modifier to keep the sharpness of the corners, which made it go from a simple 1,000 faces to a whopping 80,800, which is absolutely insane for a single part of most assets. The following shows what the result is:

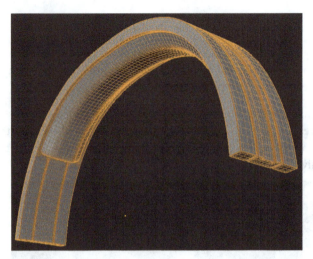

Figure 2.9 – Headband with the bevel and subdivision surface modifiers applied

As you can see, the situation isn't good at all, but the fix is simple in this case. In the **Modifiers** tab, select **Add modifier** and add the **Decimate** modifier, which looks like this:

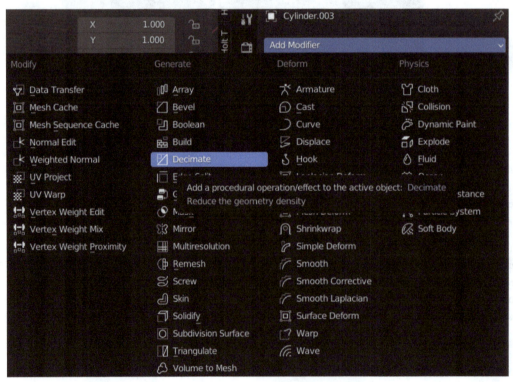

Figure 2.10 – Adding the Decimate modifier

Figure 2.11 – The Decimate modifier

There are three decimation modes, and all of them are designed to reduce the number of vertices, but they do it in different ways.

Collapse decimation

To better see what we're doing before we start, we need to go out of the edit mode (the geometry affected by the modifier doesn't show up in edit mode) and turn on the **Wireframe** option from the **Viewport Overlays** menu in the top right corner:

Figure 2.12 – Enabling the Wireframe overlay

Ok, now that we can see the edges and faces in object mode, we can change the settings in the **Decimate** modifier.

The **Collapse** mode will uniformly merge the closest vertices throughout the whole model, which can get you some funky results, but it keeps the shapes pretty well. As you drag the **Ratio** slider down, it will start merging the vertices:

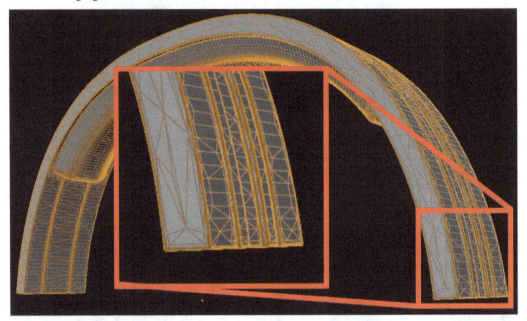

Figure 2.13 – 50% decimation using the Collapse mode set to a 0.5 ratio

We immediately notice that while the face amount was indeed reduced to half of what we had in the original, the topology has become a mess, with irregular triangles all over the model, making texturing way harder. Therefore, we won't stick with this decimation mode for this model, and there's no need to test further.

Planar decimation

This method is the third option contained in the **Decimate** modifier, and it detects faces that should be a single flat surface, keeping their outer edges and dissolving all the edges in between, leaving the ones that are necessary to keep the shape. This mode starts to take place as soon as you select it, but you can adjust the **Angle limit** slider to get different results. This is what the default **5°** setting gives us:

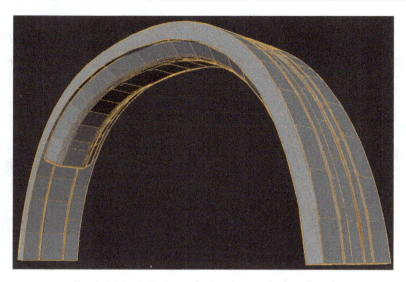

Figure 2.14 – A 5° planar decimation on the headband

At first glance, the result might look clean, but upon closer inspection, it becomes clear that this decimation won't fit our needs.

While it kept the shape pretty well, just like the **Collapse** method, it made a mess of our geometry and decimated the faces in a way that resulted in a mesh composed mostly, if not entirely, of N-gons, which are any faces with five vertices or more (see *Figure 2.15* and *Figure 2.16*):

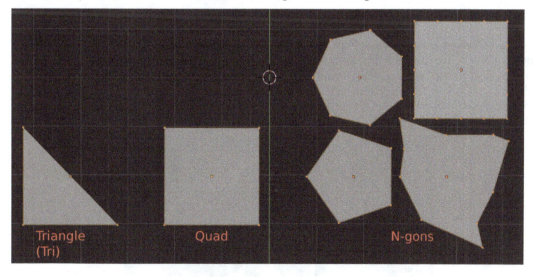

Figure 2.15 – Different types of faces

Figure 2.16 – Examples of N-gons created by the planar decimation

As you can see, this decimation method didn't work as well, as ideally, we should try to keep our model composed of quads as much as possible. Even though a few tris or N-gons here and there won't make much difference for static objects, a model composed mainly of big N-gons will undoubtedly be a pain to deal with and render later on, making the model useless in most cases.

All right, this method didn't work well, which leaves us with the last one: **Un-subdivide.**

Un-subdivide

One option left within this modifier is to **Un-subdivide** the mesh.

This option looks for quads throughout our model, joins their diagonals, and dissolves the original quads' borders, which makes this mode work especially well in models composed mainly of quads, such as our original, subdivided headband with nearly 80,000 faces. How convenient!

When we select the **Un-subdivide** option in the **Decimate** modifier, we're presented with a single **Iterations** setting:

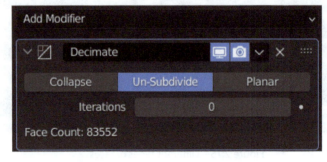

Figure 2.17 – The Un-subdivide decimation settings

When we add our first iteration, it might seem like it made a mess, just like the other methods, as it looks like it rotated all the faces:

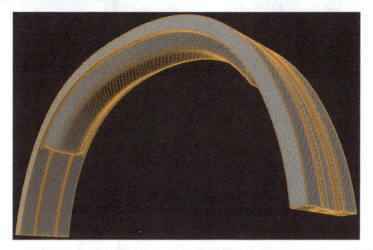

Figure 2.18 – First level of un-subdivision using the Decimate modifier

Remember, though, it only joined the diagonals of the quads and dissolved the intersections that were left, so if we add another iteration, it will join these diagonals and dissolve the edges present in the first level, creating a geometry that looks like this:

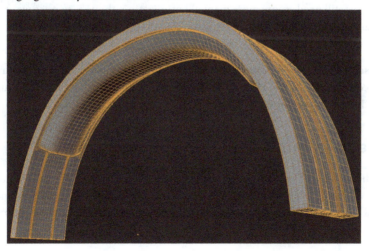

Figure 2.19 – Second level of un-subdivision using the Decimate modifier

This looks much cleaner than the other results, and if we look at our face count now, we're met with around 20,000 faces, which is still too much, but it means less manual work to do by ourselves. We can increase this number further still.

As we've seen, this tends to work better with even numbers, so we're going to increase the iterations to 4:

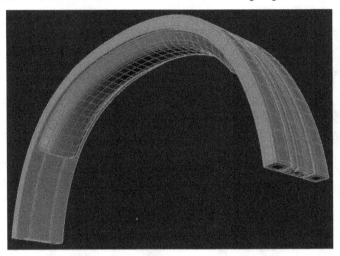

Figure 2.20 – Fourth level of un-subdivision using the Decimate modifier

Now, this looks much cleaner than it originally did, and it only has around 5,500 faces. If we try to increase more, we start to lose detail around the rounded corners, which will become progressively sharper as the decimations happen uniformly throughout the whole model, and irregular triangles might appear throughout areas that were composed of a straight line of quads in the fourth level of un-subdivision. This still needs some more manual optimization, though, so once we are happy with how the un-subdivision turned out, we can apply the **Decimate** modifier by hovering our mouse over it in the modifiers tab and pressing *Ctrl + A*.

We'll have to do some manual work by using the same technique we used on the earcups and cushions: dissolving unnecessary edge loops. How do we know what edge loops to dissolve on an object that has just been optimized by a computer, though? Well, we need to have a good eye.

What to keep in mind while optimizing a model

While Blender did a good job of optimizing our initial 80,000 faces to 5,000, we still have room for improvement, as we purposefully told Blender to stop at the fourth iteration on the **Decimate** modifier so our model still looks the same. For that, we need to see what specific areas can be optimized further.

Flat areas

If you have any surfaces that are supposed to look flat, the edges could be dissolved into a single face, so if you have anything flat in your 3D model and just applied **Un-subdivide** to it, you might want to look for some of those flat surfaces because Blender might have left a few unnecessary faces there.

Our headband has many flat surfaces, and the **Decimate** modifier left faces in all of them. They can all be dissolved using *Ctrl* + *X*. Pay attention, though, as dissolving an edge loop can cause some errors when keeping the vertices connected. Here are some edge loops we could dissolve manually in edit mode:

Figure 2.21 – More exposed loops we can dissolve

Figure 2.22 – Less exposed edge loops we could dissolve

As you can see, all of those surfaces are supposed to be flat, but the automatic decimation left them with a significant number of excess faces, which, when dissolved, takes our model to around 2,000 faces, much closer than our target of around half of that. This is what it looks like now:

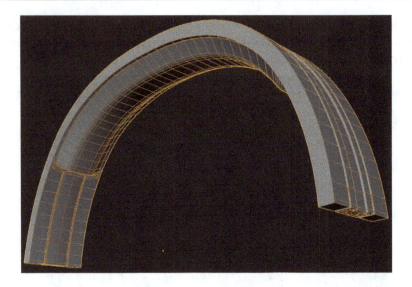

Figure 2.23 – The headband with edge loops manually dissolved after Un-subdivide has been applied

This looks great so far, but we can take it even further by taking care of the bad geometry left behind, such as stretched triangles and excessive N-gons still present in some areas.

Bad geometry

We can't deal with bad geometry using the loop dissolving technique; these faces don't allow for loops to be formed or placed, as there's no *flow* in the geometry for this loop to follow, which is only possible in quads.

The method for getting rid of bad geometry is quite similar to manual optimization. Still, instead of selecting loops, we will select each edge by hand and dissolve it, possibly combining this with other approaches, such as merging vertices into one and joining vertices with an edge to make the geometry look less messy. Messy geometry makes any object much harder to unwrap later on when we have to texture the asset.

Currently, we have a total of two areas with bad geometry, as seen in the following screenshot:

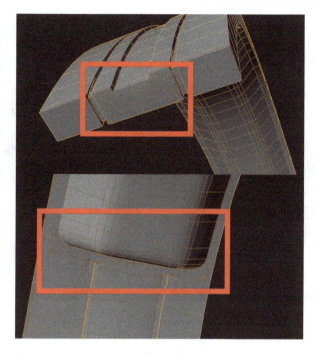

Figure 2.24 – Areas on the headband with bad geometry

Since our headphones are symmetrical, there are actually four areas with bad geometry, but we can delete half of it and work on just one side using the mirror modifier set to the *y* axis.

The main tools we'll use in this process are the *Join* (shortcut: *J*), *Merge* (shortcut: *M*), and the already well-known *Dissolve* tools. Some of them are accessible only while using vertex selection in edit mode. We'll start with the easiest part: the bottom part since there's little to no curve to worry about.

The bottom part

As this is supposed to be a flat surface, we can just leave it as a single-faced N-gon since this part doesn't make much difference in the model, and it won't be very visible since it's a connecting point between the two parts of the headband.

To start, we'll select every inner edge that composes the bottom part and dissolve them (to make the selection process easier, you can go into the bottom orthographic view using the *Ctrl + 7* shortcut (on the numpad):

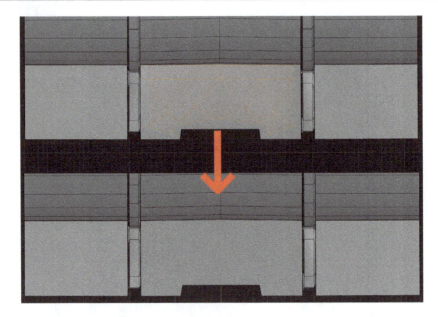

Figure 2.25 – Bottom faces turned into an N-gon

It's good practice to check the surrounding area to make sure that there's no bad geometry left by our **Decimate** modifier there too, and sure enough, there is:

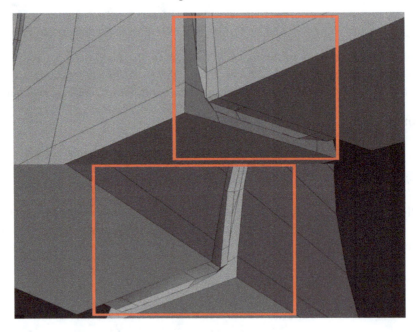

Figure 2.26 – Bad geometry around the N-gon

That's trickier to take care of but still easily achievable. In this case, we essentially have to dissolve enough edges to leave it with two faces, forming a 90° angle with each other.

This is a somewhat trial and error process, and it's impossible to tell with precision exactly which edges to dissolve since the geometry left can vary for everyone's models, but they're all surrounding that 90° angle, so you should find it, and dissolve all the unnecessary edges around it. Here are all the edges that need to be dissolved in this case:

Figure 2.27 – Edges that needed to be dissolved to correct the area's geometry

As we can see, there was much to be dissolved, including two edge loops connected to the area with bad geometry.

Upon dissolving those edges, we're left with a much cleaner 90° angle, but an N-gon was formed where we should have proper geometry and, above all, where it will certainly make a difference when unwrapping (which will be explained in the next chapter):

Figure 2.28 – An N-gon left from the manual optimization in the area

We'll need to create quads by cutting the N-gon into smaller sections. To do that, we'll use the **Join** tool, which is accessible in edit mode in Vertex select mode only.

To use it, we need to select the vertices we want to connect with an edge and press the *J* shortcut on our keyboard; then, you'll see an edge connecting them:

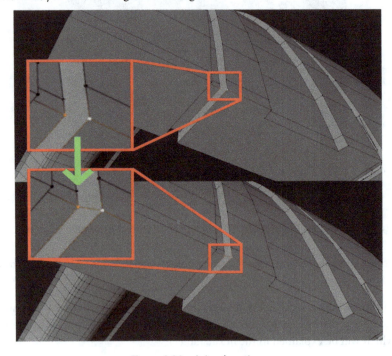

Figure 2.29 – Joined vertices

Now, we'll join the rest of them the same way we joined these first two.

A good tip is to look for edges already there and remember that Blender will not change the surface's shape but rather find the quickest path between the vertices you selected and cut the mesh's existing surface accordingly.

Here is the final result of the N-gon when cut into quads using the *Join* tool:

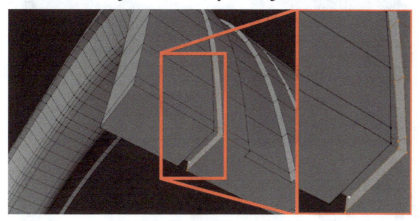

Figure 2.30 – The N-gon cut into quads

It's important to note that if your model is symmetric, you can delete half of it and add a mirror modifier to work only in half (if you don't have it enabled already).

Now that we have proper geometry in that area, we can move on to the cushions.

Around the cushions

This part needs some optimizing as well since its current state has many irregular triangles and quads, but some of them can be merged into one since they're close together:

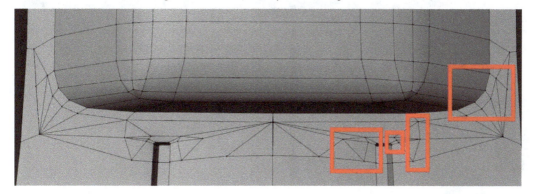

Figure 2.31 – Vertices that are close together

As you can see, most of these vertices can be merged into one, so we will combine the **Merge** tool with the **Join** and **Dissolve** tools.

To merge two or more vertices, select them (the last vertex is the one you'd want to merge the rest into) and press the *M* shortcut; you'll be presented with six options:

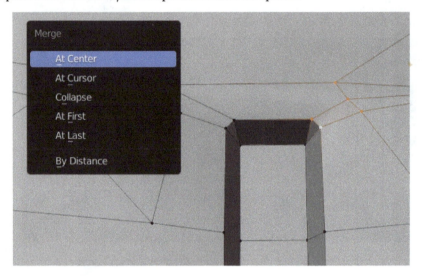

Figure 2.32 – The Merge menu

The options are pretty self-explanatory, and we're going for the **At Last** option; that's why we selected one specific vertex, which is called the *active element*, which for vertices is highlighted in white instead of orange (as we can see in the preceding screenshot). This is what you should see after merging the vertices:

Figure 2.33 – Merged vertices

Now we merge, join, and dissolve edges until we're satisfied with the result and arrive at a less messy geometry in the area. It's important to inspect your model closely, as some vertices are really close together, making it seem like there's only one vertex there.

Also, looking inside your model can help, as some vertices may even be inside your mesh. Yes, inside.

To check the inside of the mesh, we need to enter the **Wireframe** mode. To access the **Wireframe** mode, you can click the *wireframe* icon in the top right corner or simply press *Z* on your keyboard:

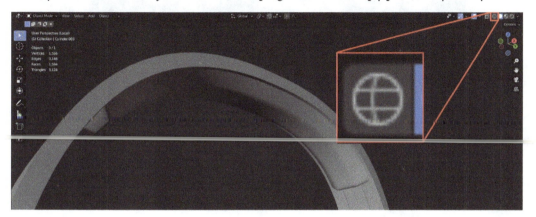

Figure 2.34 – The Wireframe mode button

Upon entering the **Wireframe** mode and inspecting the area, we can find one internal vertex, which would definitely give us trouble later on when unwrapping:

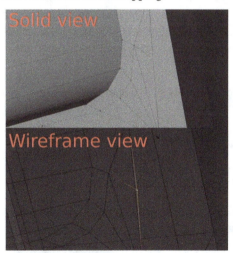

Figure 2.35 – An internal vertex

We can merge that vertex to the nearest one that makes part of our surface.

There are plenty of final results that look way less messy, so here's a good one that will be used as a final result:

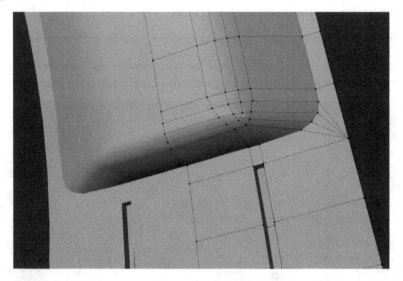

Figure 2.36 – Final optimization around the headband's rounded corners

This looks much cleaner and was made using only the **Merge**, **Join**, and **Dissolve** tools. It has only three triangles per corner, and they won't make much of a difference. The majority of the irregularities were turned into regular quads that can support loops and maintain the flow of the mesh. After this last optimization, the final face count decreased to about 1,300. And if you're wondering whether just re-modeling it from scratch would be quicker: yes, it would. Sometimes just starting from scratch is the best option if you are on a tight schedule, but it's also important to have the technical knowledge to optimize a model in case remaking it isn't an option.

This result is suitable for animation and still renders, but games could require further optimization. We'll have to reduce the polycount even more. Yes, that means we'll be getting rid of most of the detail we added, but don't worry, we'll use some computer magic to put it back without affecting the number of faces (for this to be possible, though, you need to have a copy of the high poly version, which was advised at the beginning of the chapter).

Further optimization

Currently, we have about 4,000 faces on our model, which is fine for pre-rendered projects and scenes, but it may be too much for a game, considering the size of the headphones themselves.

Along with the fact that we'll transfer the detail from the high-poly model into the low-poly one (that's why we kept a copy of the high-poly version from the beginning of the chapter), we'll dissolve the edges that compose all the details we added by selecting all the faces that compose them, deleting them, and filling the holes with new faces. Do this for all the parts that can have less detail.

One good tip is to keep the silhouette in mind and not dissolve as much geometry in the headband, for example, as you would in something such as the cushions and earcups, which don't make up much of the headphone's silhouette.

As you already have the technical knowledge to perform this optimization, we'll only add pictures of the before and after to demonstrate what kind of things can/should be dissolved (the objects were isolated with / *(forward slash)* for better viewing):

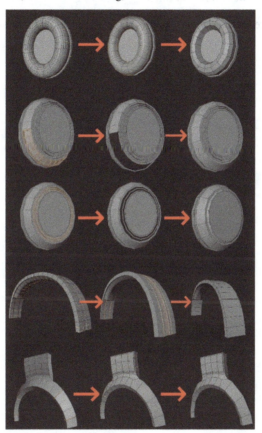

Figure 2.37 – Further optimizations that can be made if the model is supposed to work in real time

All the optimizations were made using the same tools and techniques used throughout the chapter.

As you can see, all the smaller details were deleted in favor of having fewer faces, which means less memory is required to render it. With this optimization, we got the number of faces down from 4,000 to around 840, which is much easier to run inside a game.

So, how do we get all the details back? We will bake those details back using textures, but that requires a few steps that will be covered in the next two chapters, starting with UV unwrapping.

Summary

In this chapter, we covered why having an optimized mesh is important both for running it in a game engine and for facilitating the next steps, as well as learned the most used tools to do so, such as the **Join**, **Dissolve**, and **Merge** tools, which can be used in combination with the **Decimate** modifier to create optimal results that could be used for any purpose. We also covered further optimization in case the model is intended to be run in real time.

Now, we're ready to prepare these headphones for texturing, and thanks to our optimization, it will be much easier!

3
UV Unwrapping Your 3D Asset

With our headphones fully optimized, we're now ready to prepare them for texturing through a process called **UV unwrapping**, which basically consists of placing cuts (seams) at strategic places on your mesh in order to lay all the polygons flat, in a 1:1 image called a **UV map**. This will allow us to texture our asset smoothly, without worrying about stretched textures.

This chapter will cover the following topics:

- How to place seams in the correct places based on the shape of the mesh
- How to unwrap our asset
- How to use the UV Editing workspace
- How to check for stretching in the textures
- How to minimize stretching in the textures
- How to pack the object's UVs

It's important to remember that, in this chapter, we will unwrap the high-poly headphones in order to demonstrate more complex techniques, but if you plan on making this a game asset and you made a low-poly version as well, the high-poly version doesn't need fully working UVs (it does need UVs, though, but you can use automatic UV unwrapping methods, which will be mentioned during the chapter). The logic presented here still applies to the low-poly version, so you can use it to unwrap your low-poly model as well.

Finding the right edges on which to place seams

The first step to unwrapping any asset is the seam placement on the edges of our mesh, but not all edges. We need to place just enough seams to lay our object's geometry flat, quite literally like a carpet. There's no *correct* result; as long as we're satisfied with the outcome and stretching is minimized, it can be considered *correct*. Remember, though, that those seams might be visible in the textures since the parts will be separated, so we'd want to use as few of them as possible, and in places where they won't be seen as much.

So, where do we put seams? That depends entirely on the shape of your mesh. But generally, seams can be put on edges that separate different surfaces, which are generally the edges with sharper angles, although some more seams can be placed elsewhere to get rid of stretching. Let's see what this means through examples, using a few of the main primitive shapes available.

Cube

We all know the most popular cube unwrap: the cross. But how do we achieve it? To understand better, let's look at the cross first.

For demonstration purposes, a physical flat cross will be modeled to represent the final result, but the actual result won't have to be physically modeled.

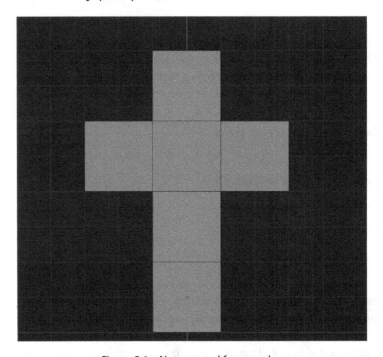

Figure 3.1 – Nets created from a cube

By looking at the preceding figure, two conclusions can be made: all the borders represent cuts (seams), and there is a long string of faces without any seams in it, which may seem odd given that all six faces of the cube have sharp edges. But remember, the objective is to place as few seams as possible, so a long string of consecutive faces can be considered one surface. What about the two faces to the sides? Let's see why they're there by bending the faces almost back together (the long string of faces will be marked in blue, and the edges of those two side faces will be red):

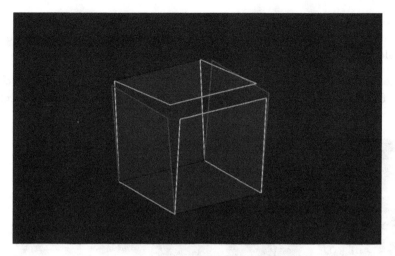

Figure 3.2 – Cube almost bent back together

Now, we can more clearly see why those side faces have seams: they were holding our main string of faces together, making it impossible for the unwrapping to happen. If the seams on the sides were not there, the net would be like this, as seen from above:

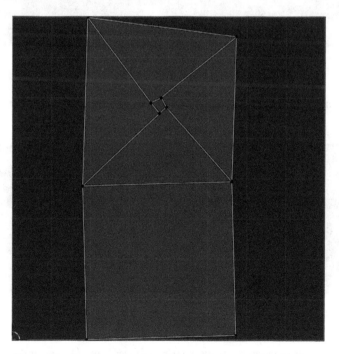

Figure 3.3 – Unwrap of a cube, without the cuts on the side faces

This unwrap has two fatal flaws: it overlaps itself (which makes texturing for anything outside Blender impossible), and the faces around that central face are stretched to such an extent that no texture in this world would look decent. It basically breaks the laws of physics.

Coming back to the cross unwrap, if we were to leave only the borders marked and finish closing this cube, we'd be left with a perfect cube with its seams where the parts from the unwrap meet (marked in red):

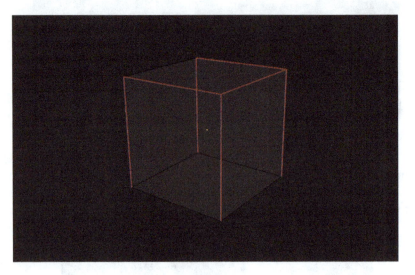

Figure 3.4 – Seams in a cube to form the "cross" unwrap

There are a few different unwraps for a cube that will work nicely. Here's another suggestion that works well:

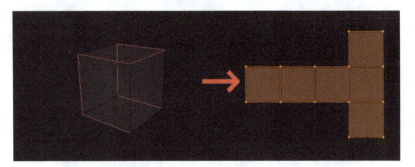

Figure 3.5 – Alternative unwrap for a cube

Notice how the logic remains the same: the main string of faces is maintained and the faces that were holding this string together were cut. Now, let's see how that logic applies to the other primitive shapes.

Cylinder

For a cylinder, the logic is still mostly the same, with the same long string of faces going down the middle, except this time, there are no sharp corners going down the sides. In such cases, we'll need to place only one cut in the long string instead of two or four, as in the cube. Again, by separating the faces that hold the main faces together, we get the following:

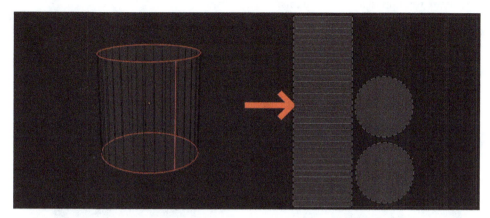

Figure 3.6 – Unwrap of a cylinder

You may have noticed that the top and bottom faces were separated completely (in what are called **UV islands**), unlike the cube. That's because if we leave one edge connected, the textures will have no seams in that face only, making it look a bit weird since all the other edges will have visible seams. Plus, the UVs will come out diagonally, which will cause pixelation problems in straight lines when applying lower-resolution textures:

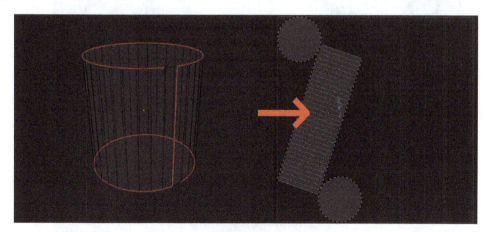

Figure 3.7 – Cylinder unwrap with one edge left without seams on each side

Yes, this is a perfectly acceptable unwrap and there's no stretching, but you'd have to manually manipulate it to straighten it, so why bother doing this when we have a perfectly straight map right out of the box?

Torus

The torus can seem pretty confusing to unwrap at first glance, as it has no sharp edges whatsoever. So how on earth are we supposed to unwrap this?

Well, if we think about it, a torus looks like a long cylinder bent at a 360° angle, so the unwrap can be done in almost the same way as a cylinder:

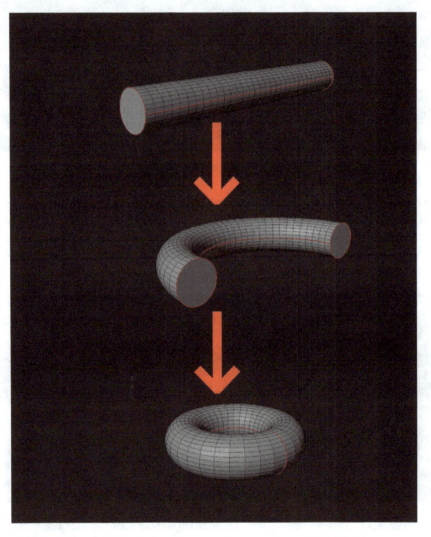

Figure 3.8 – Torus seam placement in comparison with a cylinder

Now, it becomes much clearer where to put the seams in a torus and why it works. By placing one seam horizontally and one vertically, we get this result:

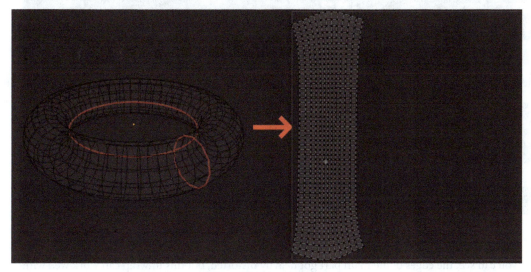

Figure 3.9 – Torus unwrap

This unwrap is good and has little stretching but, as you can see, the outer edges got a bit distorted, which will make the textures behave weirdly around the corners. That means this UV island requires manipulation. It's worth noting, though, that not all UVs with slightly crooked edges will need manipulation; this one requires it because it's an object that is flat in at least one axis, so to say. In this case, the UVs have to reflect that.

UV manipulation will be covered in more detail when we go to unwrap our headphones.

UV sphere

A sphere can be tricky to deal with as well since it also doesn't have any sharp edges, so there's no way to get the stretching down to absolutely zero. There are ways to minimize it considerably, though.

One way is to unwrap it like a cylinder, just like the torus, but this seam placement would need the same kind of manipulation that a torus would:

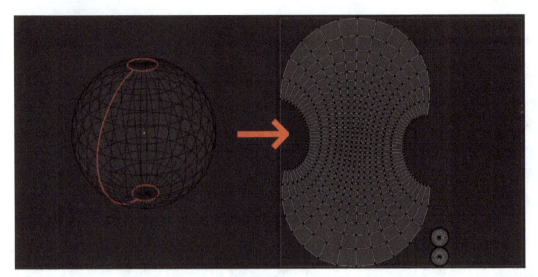

Figure 3.10 – UV sphere unwrapped like a cylinder

As you can see, the edges of the main part got heavily distorted, but this unwrap can work well with said manipulation.

Another unwrap that has little distortion but doesn't need manipulation to work right is to use four seams, cutting the sphere into four different parts:

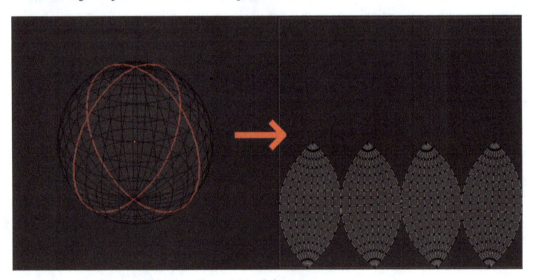

Figure 3.11 – Sphere unwrap

Note that there are four seams now, so if you come across a case where your model is spherical, you'd have to decide whether four seams will compromise the quality or not – that is, if you can't hide them in any way.

This might seem like a distorted result, but if we check for stretching, we'll see that there's actually minimal stretching all around the sphere (the closer to dark blue, the less stretching, and the closer to yellow, the greater the stretching):

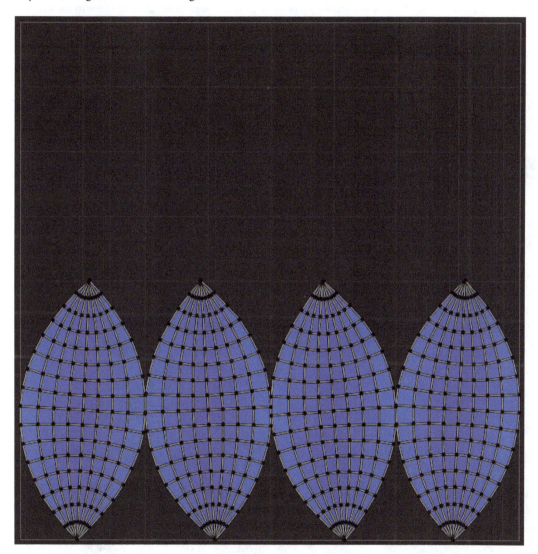

Figure 3.12 – UV sphere "four parts unwrap" stretching displayed

Alright, now that we covered the main basic shapes and where to cut them, we're ready to unwrap our headphones!

Unwrapping the headphones

To start, we need to try and visualize the shapes we covered in our model and apply the same logic to them – that is, maintain the most visible parts without seams and avoid placing them in the main string of faces.

In order to see what our UV unwrap looks like, we need to access the dedicated tab for this: the **UV Editing** tab. To enter it, we have two options. The first one is to select it in the tabs above the viewport:

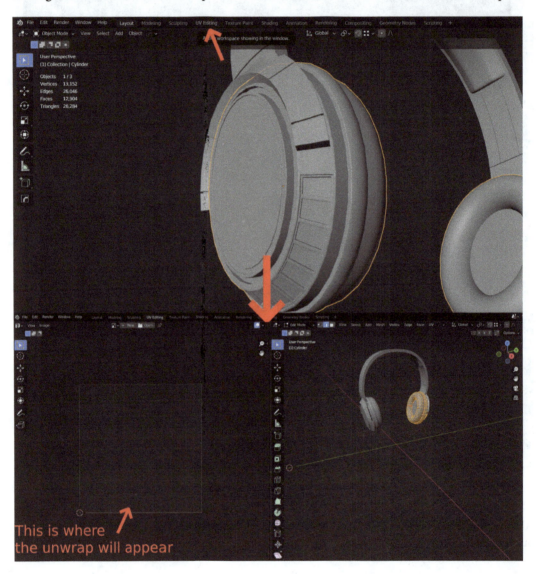

Figure 3.13 – Entering the UV Editing workspace using the tabs

Another way to enter the UV workspace is to right-click on one edge of the viewport tab, select either **Vertical split** or **Horizontal click**, then click again where you'd like the viewport tab window to be split (you can also join or swap areas using this method). Then, we should select the *Editor Type* icon in the top-left corner and finally, select **UV Editor** from the drop-down menu:

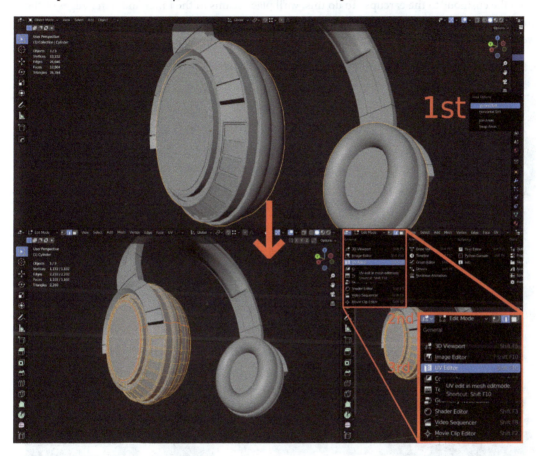

Figure 3.14 – Alternative way to enter the UV Editor

This workspace is made to display and tweak your UV maps, as well as to test for stretching in it. To tweak it, you can use the same main tools and their shortcuts from the viewport's **Edit Mode**, such as **Grab** (*G*), **Scale** (*S*), and **Rotate** (*R*). Though, to see the full unwrap, we have to select all the faces/edges in **Edit Mode** using the shortcut *A*.

We'll start with the earcups and cushions.

Cushions and earcups

When we made our earcups, we decided to keep the two different parts as one, but now, we need to place seams in between them in order to get a good unwrap, since the overall shape changes drastically from the cushions to the earcups. To do this, we'll place seams in the inner and outer edges of the cushions by selecting them in **Edit Mode**: right-click and select **Mark Seam**. You should see the edges we selected earlier turn red; this means that those edges now represent a seam:

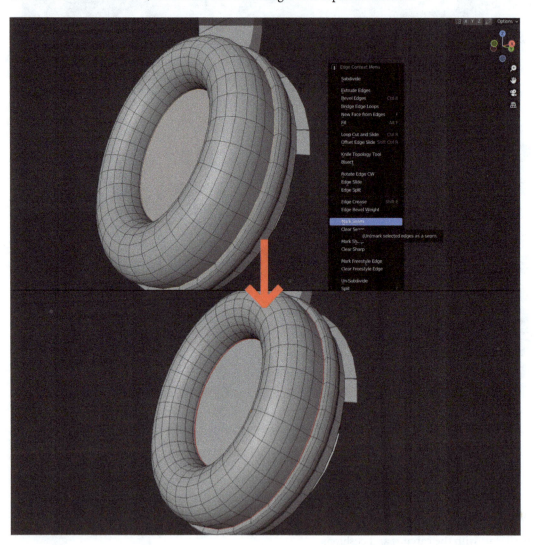

Figure 3.15 – Marking seams to isolate the cushions in a UV island

Now, with the cushions separated, we notice that it resembles a torus, so we'll proceed to make a cut going vertically in an area where it wouldn't be seen with much frequency – in this case, the back part:

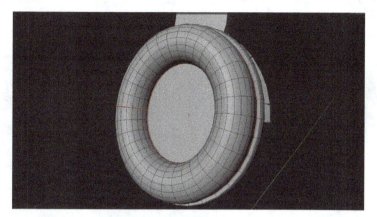

Figure 3.16 – Seam placed in the back part of the cushions

With this, the cushions are ready to unwrap, and we can move on to the earcups before telling Blender to unwrap them fully.

They were made from a cylinder, and we kept their main shape like a cylinder, so that means we'd have to place a cut in the edges and another one going down the sides. For the loop going down the earcup's side, though, it's a good idea to place it in the same edge loop as the seam going down the cushions:

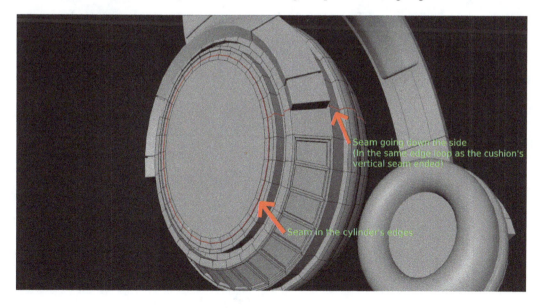

Figure 3.17 – Seam placement in the earcups

It looks like we'd get a successful unwrap at this point, so it's time to see how it unwraps.

To unwrap our model, we need to select every face, edge, or vertex using the *A* shortcut, then press the *U* shortcut to open the **UV Mapping** menu, where you'll be presented with several options. We're going to select the **Unwrap** option:

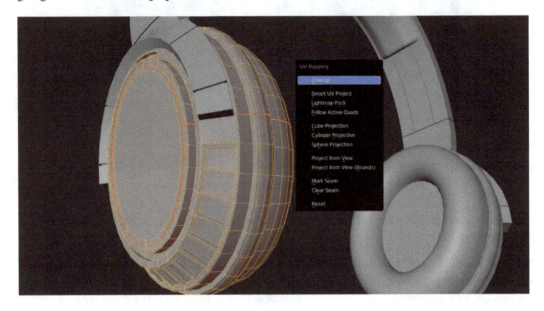

Figure 3.18 – UV Mapping menu

When we select the **Unwrap** option, our unwrapping result will appear in the **UV Editing** workspace.

Tweaking the UVs

Now, with the UV editor open, the result for our unwrap looks like this:

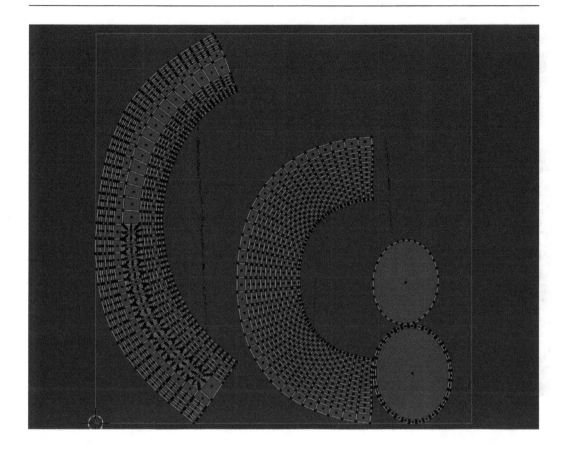

Figure 3.19 – Earcups and cushions unwrapped

As you can see, the object was unwrapped, and the result is bent due to the round shape, but since it has a *straight flow of edges* in the 3D workspace, we'll need to straighten it. To avoid trouble while straightening it, let's check for stretching to see whether there are any areas we should worry about or add extra seams to.

To do this, we need to select the **Show overlays** menu in the tab's top-right corner and then check **Display Stretch**; the unwrap should turn dark blue where there's no stretching and yellow or green when there's a lot:

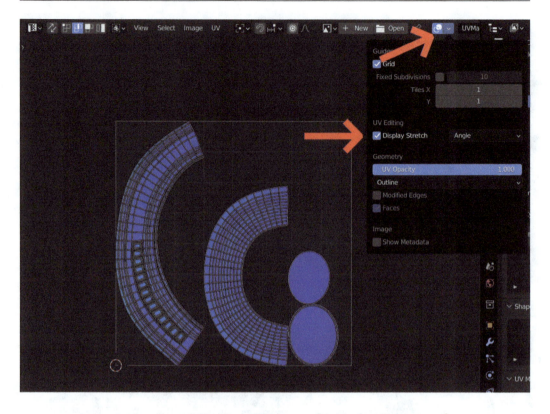

Figure 3.20 – Displaying the stretching in the UVs

At first, it may look like there's not much to worry about, but upon closer inspection, we notice something that was covered by the edges close together: we have a lot of stretching around the holes we carved for detail.

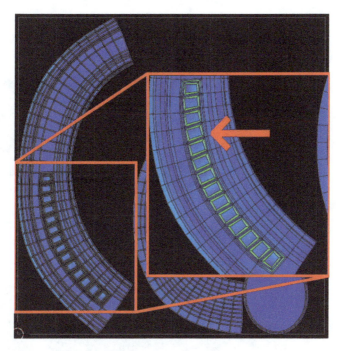

Figure 3.21 – Stretching around the detail

In order to get rid of the stretching, we're going to apply the same *main face string* logic, starting by isolating this area in its own separate UV island, so we have to place seams around the edges of the hole we carved in and unwrap it again:

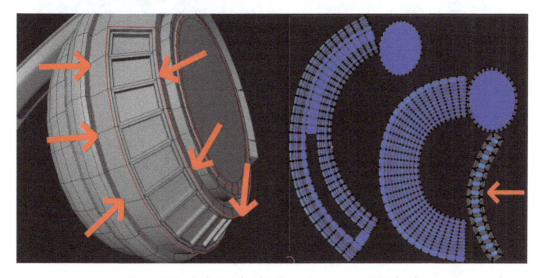

Figure 3.22 – Isolating the detail area in a separate UV island

The stretching on the piece it was attached to disappeared but it remained on the detail itself, so now it's time to apply the actual face string logic here. In our case, it would be the middle faces, which contain all the holes (marked in orange), while the outside edges are not part of the face string:

Figure 3.23 – The long string of faces of the detail

Okay, now we have to figure out which edges are holding this string together in a distorted way and apply seams to them. Let's have a look at our UV island:

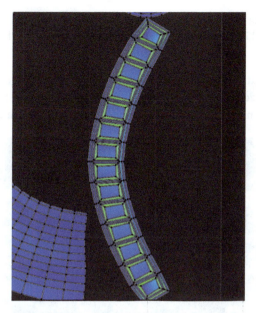

Figure 3.24 – Isolated UV island with stretching

By looking at it and taking into account the shape of the actual detail, it seems like the diagonal edges coming out from the central, bigger faces are what's holding together the main string of faces in a way that makes it appear stretched. (Remember: if you don't place seams on an edge, it's going to stay connected, even if Blender has to stretch those edges/faces to do so.) So, we have to select all those diagonal edges and mark them as seams, then unwrap the earcup again using the *U* shortcut and select **Unwrap** afterward:

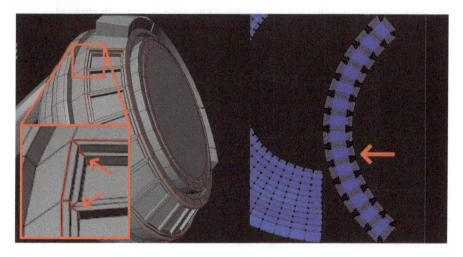

Figure 3.25 – Seams placed on the diagonal edges in order to reduce the stretching

As we can see, the stretching was completely removed from the detail, and now we can move on to straightening the other two UV islands, which is incredibly simple since we'll let Blender do the hard work for us.

To start, we choose one quad – any quad in a UV island. After that, in **Edge** select mode, select one of the edges and scale it by zero on the axis to which this edge is the closest to being aligned (don't worry about the stretching in the faces around it; Blender will deal with it later):

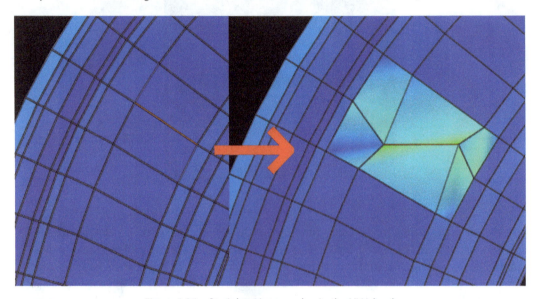

Figure 3.26 – Straightening an edge in the UV island

Then, you do the same with the other three edges, until you have a quad with every edge aligned in a straight manner:

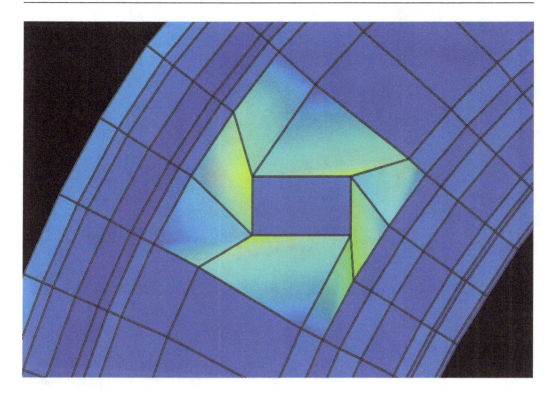

Figure 3.27 – Straightened quad in the UV island

Now, we want to select it in **Face** select mode, then select all the island's faces by hovering your mouse over the island, pressing *L*, and right-clicking afterward; then, select **Follow Active Quads** from the menu that pops up, and just like that, our UV island will get straightened automatically since we told Blender to apply that same straightened shape to all faces in the UV island.

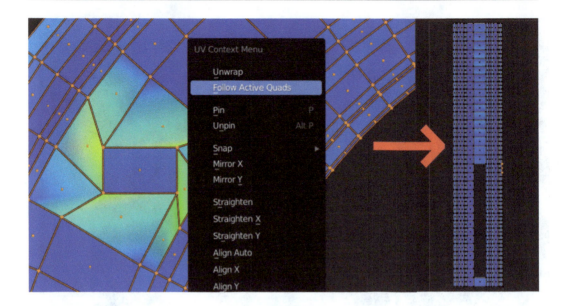

Figure 3.28 – Straightened UV island

Great, now we can do the same with the other islands that need to be straightened; rearrange them in a more organized layout (maybe even some manual UV tweaking afterward), and we should have something like this:

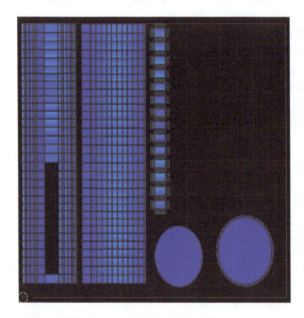

Figure 3.29 – Ear cups and cushions fully unwrapped

There may be minimal stretching left; it doesn't have to be perfect, but rather good enough. This looks very good, with mostly blue faces, and everything is straight. It's also good to note that there is more than one way to unwrap our asset; we have shown only one approach. We're now ready to move on to the headband.

Headband

For this step, it's good to apply any modifier you might have, except the bevel and/or subdivision modifiers (if you chose to add them for some reason).

The headband will be a little trickier since it has its cushion and triangles, and the overall shape is much like a cube. Instead, though, we'll use the same logic as for a cylinder because it is much longer than a cube, resembling a cylinder more, plus the border faces won't be visible in the final product.

We'll start by making the cuts as we would in a cylinder, separating the borders, and placing a seam going from one border to another (to make the process easier, you can isolate the object by selecting it in **Object Mode** and pressing /):

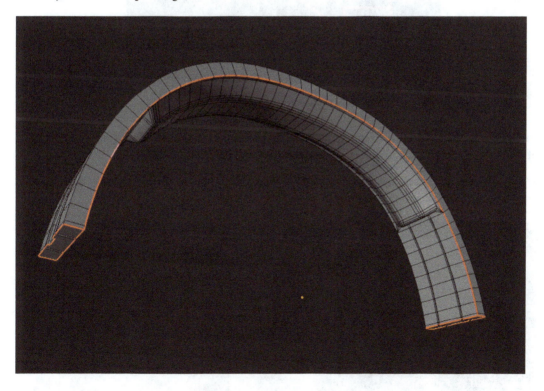

Figure 3.30 – First seams placed on the headband

Now, our unwrap should look like this:

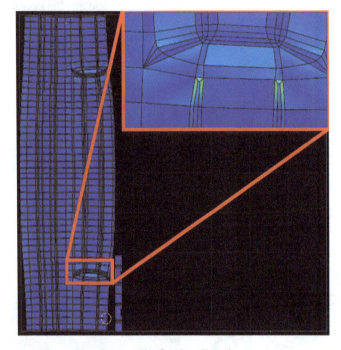

Figure 3.31 – The first headband unwrap

As we can see, the unwrap has stretching in the main part and the sides are bulging outward because of the cushions, in addition to the strong stretching around the detail we added.

We'll proceed to add seams to isolate the cushions and on the borders of the detail, since those edges are what's causing that strong stretching by holding it together:

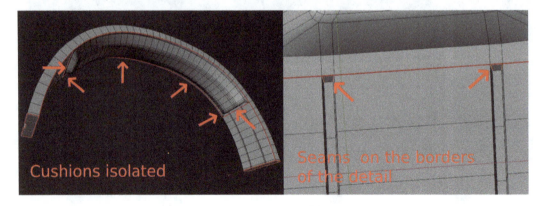

Cushions isolated

Seams on the borders of the detail

Figure 3.32 – Extra seams added

Remember that if your object is symmetrical, you should add seams to both sides. Now, the unwrap looks like this:

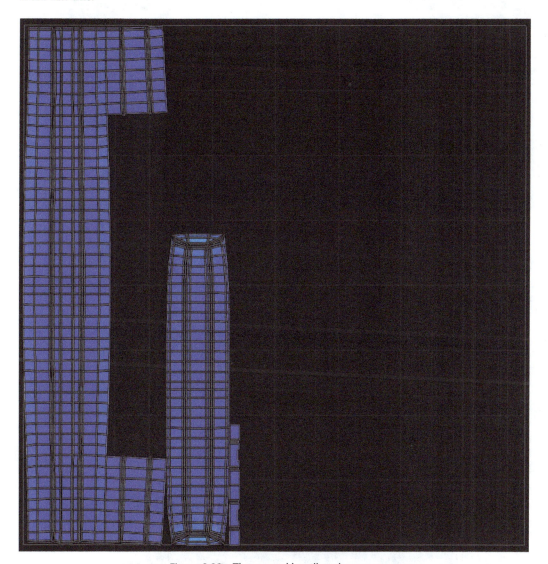

Figure 3.33 – The second headband unwrap

It's looking better, but we could improve it. The round edges of the cushion, which contain the triangles, can be isolated on another island in order to make it possible for us to straighten the middle part. This might cause seams in the texture though, so if you don't want it, you can straighten the main part and leave the rest since it's possible to work with this.

If you decide to straighten the cushion, here's how to do it: simply isolate the part with the triangles with seams on both sides, like this:

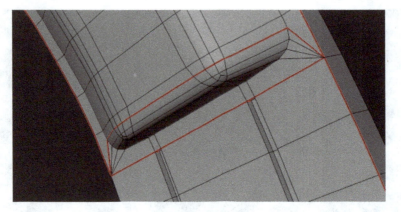

Figure 3.34 – Area with triangles isolated

This is case-specific, so it should be done only when needed, but nothing is keeping us from testing anything!

Those seams leave us with this unwrap for the headband's cushion:

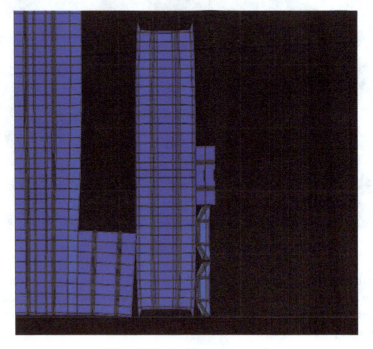

Figure 3.35 – Cushion unwrapped

We can now straighten the two main islands, maybe even tweak them a bit according to necessity, and we're left with this:

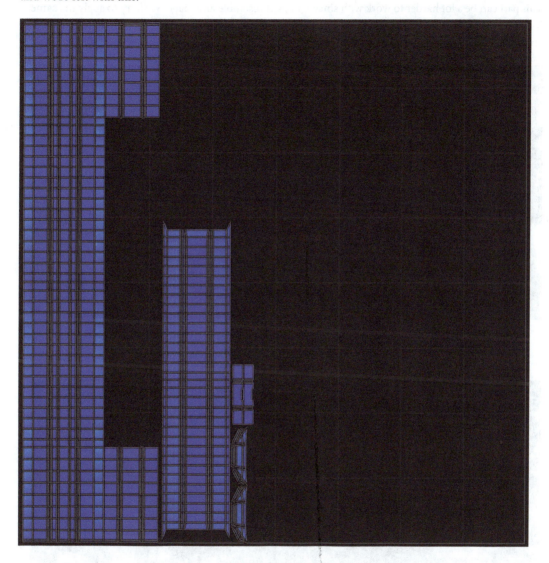

Figure 3.36 – Final headband unwrap

And, with UVs that look good and have minimal stretching, we can move on to the last piece of our headphones.

Connecting part

This part can be a lot harder to work with since it doesn't resemble any shape. We'll try to apply the same long face string here but adapted slightly. In our case, we want the larger sides to remain connected, so the main face string looks like this (marked in orange):

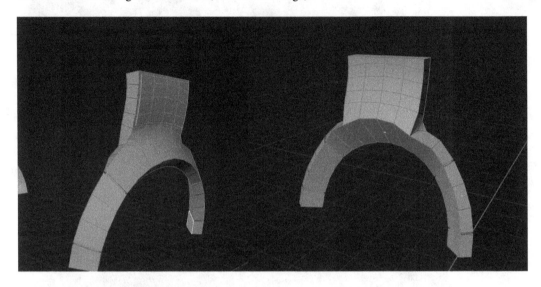

Figure 3.37 – The main string of faces in the connecting piece

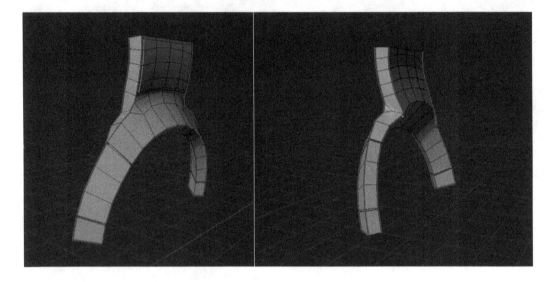

Figure 3.38 – Seams marked in the borders

For this unwrap, it's impossible to get a perfectly straight unwrap because the object itself is not very regular, so we'll do a *butterfly* unwrap, which we already got by placing these seams:

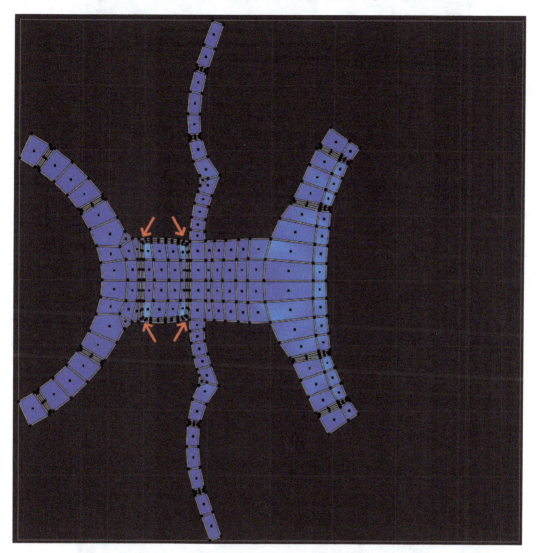

Figure 3.39 – "Butterfly" unwrap of the connecting part

This already looks nice, but there is stretching on the faces that compose the square that protrudes forward in the upper portion (highlighted with arrows on the preceding image), so we'll separate that too by placing seams around it, as well as in the edges holding it together to minimize stretching inside the protrusion itself:

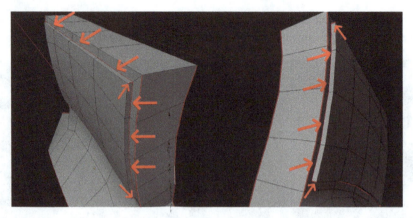

Figure 3.40 – Isolating protrusion

And we're left with an unwrap that looks like this:

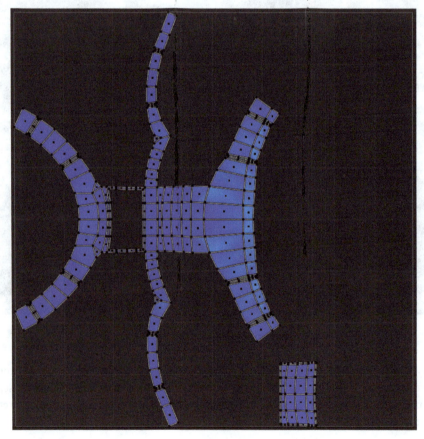

Figure 3.41 – Final "butterfly" unwrap

For a shape like this, there are fewer options for unwraps, but again, this is not the only one.

We've got the job done. There's almost no stretching in most places, and the areas that do have a bit more stretching mostly don't show up, so we can consider this successfully unwrapped. We're now ready to pack all the UVs from all the parts into a single UV map so that we can use only one texture for all parts, which makes the asset even better in terms of rendering performance.

Packing the UVs

UV packing is the process of optimizing the UV island's placement in the UV map, and it can be done to one object or several. It makes it possible to use just one texture in several objects (only necessary if you want/need one texture map for multiple objects or for baking the textures correctly, which will be explained in depth in the next chapter. You'll use this for exporting your model to another program besides Blender, for example.)

Once again, if you have a high-poly version of your model and want to transfer the details from the high-poly to the low-poly version, *only* the *low-poly* version needs fully working UVs, so you'll only want to do proper UV packing for the low-poly version.

To properly pack our UVs, we need to think about both the size and visibility of each object, and it's pretty intuitive: the bigger or more important it is, the more space it should occupy in the UV map.

You should *only* manipulate the scale, rotation, and scale of the *islands* at this point, as all the fine-tuning of the faces composing those islands should be done while unwrapping the objects themselves.

To manipulate an entire island, you can hover your cursor over an island and press *L* to select it, then use the shortcuts *G*, *S*, and *R* to grab, rotate, and scale the islands, respectively. In this process, we're supposed to fill as much space as possible with all the islands.

We can also try to let Blender pack them automatically and see how it goes. To do that, we need to select all of our objects in **Object Mode**, go into **Edit Mode** with them selected, and then select all faces with the shortcut *A*. You should now see all the UVs of all the objects at the same time, as we left them:

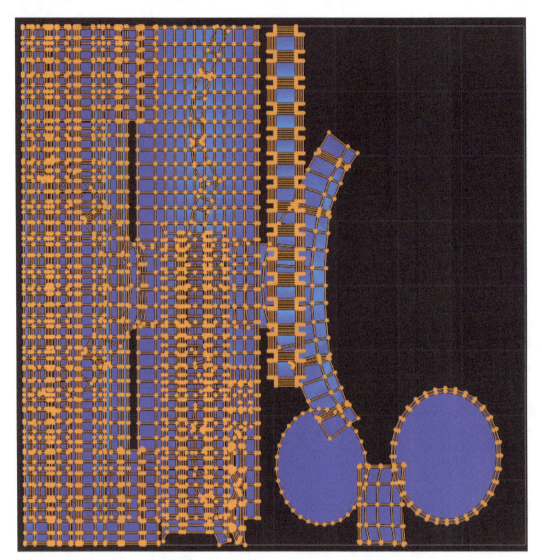

Figure 3.42 – All the UVs displayed at the same time

To make Blender try and organize this mess, we can select all the faces in the UV Editor, then click on the **UV** tab in the top-middle of the screen, and then select **Pack Islands**, and Blender will mostly give out decent results – a very good starting point at least:

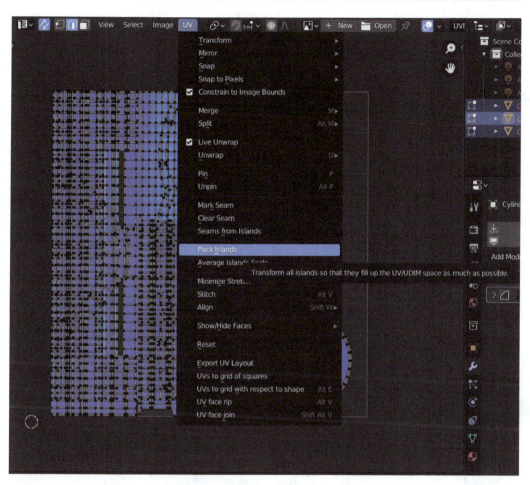

Figure 3.43 – Automatic packing option

And this is the result right out of the box:

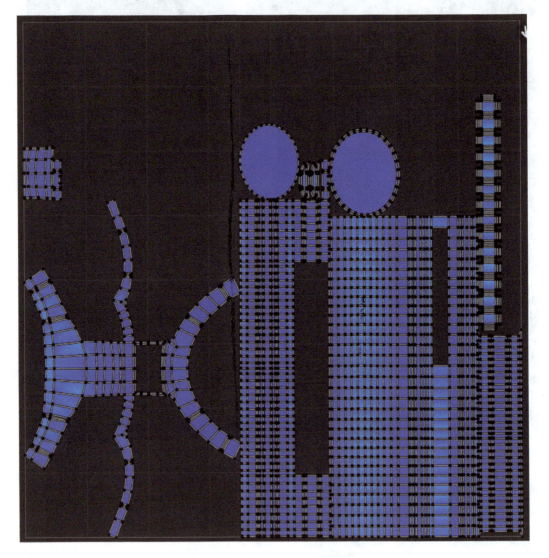

Figure 3.44 – Blender's packing result

This is a very, very good starting point, but notice how there's a lot of empty space in the upper portion: that's lost resolution. We can now start tweaking the location, scale, and rotation of the separate islands to get a better pack. There's not much else to say about the process, since everyone's object is different, so just try to position the straight islands either horizontally or vertically, not diagonally, and make the parts occupy as much space as possible, according to their importance, size, and/or visibility. We can get a pack like this:

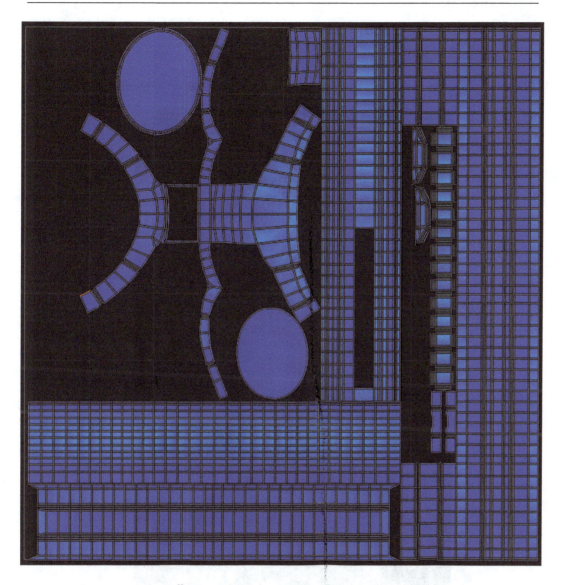

Figure 3.45 – More efficient pack, made manually using the automatic pack as a base

It's important to respect the edges of the UV map since the textures won't work on anything outside of it.

We're officially done with the unwrapping, but we can do some extra checking to see how those UVs would work in practice.

Checking for practical use

To check for practical use, we'll apply a placeholder texture in all objects.

We'll start by going into the **Materials** tab on the right side of the viewport, then click on **New**:

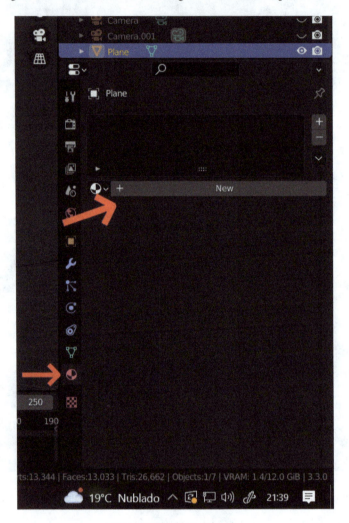

Figure 3.46 – Materials tab

Then, settings should pop up underneath it, in which we should click on the dot to the left of **Base Color**, then select **Image Texture** (texturing will be covered in much more depth in the next chapter):

Figure 3.47 – Adding an image texture

From there, we can select **New**, **Blank**, and finally, **UV Grid**:

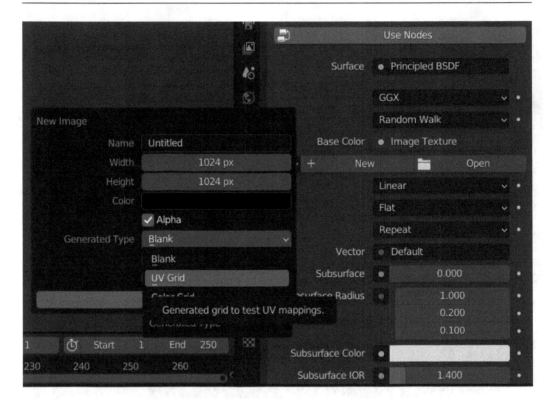

Figure 3.48 – Setting UV Grid as the texture type

When you click **OK**, the texture gets applied, and to see it, we need to go into **Material Preview Mode** on the viewport, by selecting its icon in the top-right corner:

Figure 3.49 – Material preview

Now, we can see the textures displayed on our object, and to apply the same texture to all of the objects, we select them, followed by the object with the texture, then press *Ctrl + L* and select **Link Materials**, and voilà, we have our material applied:

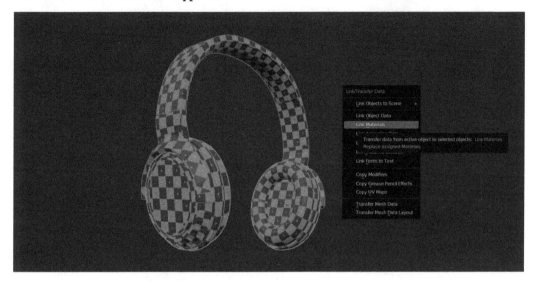

Figure 3.50 – Checkered pattern applied to all objects

Now, we should check for deformed or crooked squares, as well as overly large or small ones, given the relative dimensions of the actual objects and parts of the object. In our case, though, since our UVs were made mostly with minimizing the stretching in mind, our objects and the texture look good and don't need reworking on their UVs.

Now comes the fun part: texturing. We're finally going to bring some life into our headphones!

Regardless of whether or not you have a low-poly version of your model, the texturing will be made on the high-poly version, and those textures will be transferred to the low-poly model.

Summary

In this chapter, we covered why it is important to have proper UVs, as well as how to unwrap the primitive shapes, and how to apply similar logic to more complex shapes, making use of the seam placement found in those same primitive shapes. Efficient ways of manipulating the UVs for more optimized results were also covered in this chapter. We're now ready to texture our asset and give it some personality.

4

Texturing Your 3D Asset Using PBR and Procedural Textures

With our asset fully modeled and unwrapped, it's time to give it some color and physical properties in order to make it more believable and give it some personality. To do this, we're going to make almost exclusive use of the **Shader Editor** tab, in addition to the 3D Viewport.

In this chapter, we will cover the following:

- Different texturing workflows – PBR and procedural texturing
- The best sites to look for textures
- How to work with the free built-in **Node Wrangler** add-on

Keep in mind, though, that we'll texture a high-poly model, since it has more detail. If you plan on making game models, we'll cover how to transfer details from a high-poly model to a low-poly one at the end of the next chapter, also about texturing.

Buckle up – this will be a long ride.

Setting up your workspace

Before we get started, we need to make sure that we can edit textures; we also need to know whether all of our textures will display correctly and whether we can see the changes live as we make them.

We'll start by either opening the shading workspace or opening **Shader Editor** as explained in the previous chapter. This is so we can actually see what we're doing.

Then, we should scale the objects to the size they are in the real world (if you plan to have them at real-world scale), as often we end up modeling our assets either too big or too small. After scaling the object, make sure to apply the scale by using the *Ctrl + A* shortcut and selecting **Apply scale**. By doing this, we can ensure that our textures will be calculated at the right scale.

Finally, so we can see the textures with their physical properties displayed (roughness, bumpiness, transparency, and so on), we can enable the **Material Preview** mode in the 3D Viewport by simply clicking on the checker-patterned circle in the top-right corner:

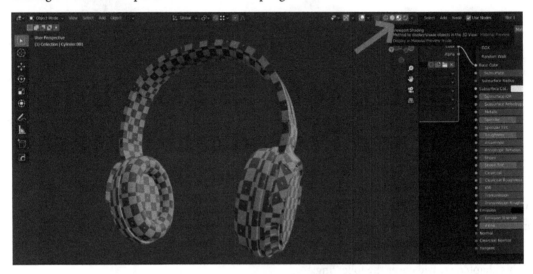

Figure 4.1 – Enabling Material Preview mode for the 3D Viewport

Now, we can remove the placeholder material we applied earlier by deleting it from the assigned materials. To do that, we select our object in Object Mode and go to the **Material Properties** tab (**1**), then we select our placeholder (**2**), and click the minus button to the right (**3**). Do this for every object:

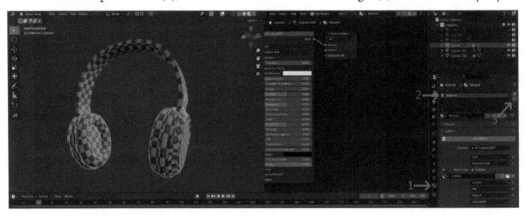

Figure 4.2 – Removing the placeholder material

Your object should be completely white now; this means it has no materials. We're ready to begin. We'll start with the cushions and earcups, since we can demonstrate most texturing methods with them.

A physically based rendering workflow

A **physically based rendering** (**PBR**) workflow consists of using textures that were designed with 3D use in mind based on real-life physical properties. It's usually the easiest way to apply textures to our models that give realistic results. We can find countless PBR textures on the internet, and some sites even offer commercial licenses to use them. We'll be using `polyhaven.com` and `ambientcg.com` for our PBR workflow.

We'll start with the cushions. Select the earcups in Object Mode, select every face that makes the cushion in Edit Mode, and then add two blank materials by clicking twice on the plus sign (**1**) above the minus sign mentioned previously (the first material gets assigned to all the faces in the object by default, so we add a second material to assign specific faces to it since our object has more than one material). After that, select the second material from top to bottom (**2**), while still in Edit Mode with the faces selected, and click **Assign** (**3**):

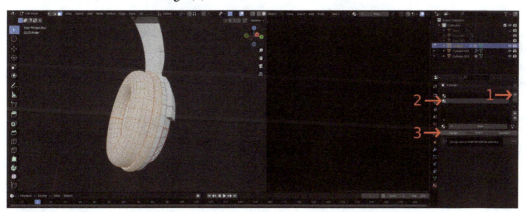

Figure 4.3 – Assigning new materials to the cushions

Now it's time to look for the texture we're going to use for the cushions.

Looking at our references, we can see that cushions are usually made out of either cloth or leather. We're going with leather, but there's nothing stopping us from testing other materials.

The chosen texture comes from `ambientcg.com` is called **Leather 027**, and looks like this:

Figure 4.4 – The Leather 027 PBR texture from ambientcg.com

After choosing the material, you can choose the resolution you want the texture in, which usually ranges from 1K to 8K or sometimes even 16K. For our case, though, anything above 2K is extremely excessive and could have an impact on performance. After choosing the resolution and format (PNG or JPG), download the texture and unzip the files into a folder. You should see all the necessary texture maps as image files:

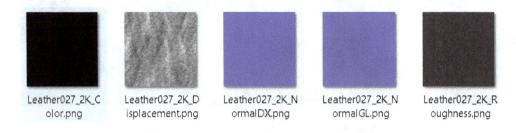

| Leather027_2K_C olor.png | Leather027_2K_D isplacement.png | Leather027_2K_N ormalDX.png | Leather027_2K_N ormalGL.png | Leather027_2K_R oughness.png |

Figure 4.5 – Texture maps downloaded and unzipped as PNG files

The reason more than one image is needed might not be obvious, though.

Different texture maps

The different maps are needed in order to give our surface more realistic physical properties, such as shininess, metalness, and bumpiness. Each one of the maps takes care of one of those aspects.

Color/diffuse map

This map, as the name implies, handles only the colors, and if we apply only the leather color to an object, it looks like this:

Figure 4.6 – The leather color map applied alone, rendered with raytracing

As you can see there's almost no detail, as the leather itself has very little color variation. This is where the other maps we downloaded come into play.

Roughness/gloss map

A roughness or gloss map is a black-and-white image that will basically tell Blender how reflective a specific area is. The darker an area is in the image, the more reflective it becomes inside Blender (for gloss maps, this is reversed, but it serves the exact same purpose). If we now apply the roughness map to our surface, we get this:

Figure 4.7 – Roughness map applied along with the color map

We can see much better now; the details are more clearly shown as this leather has a large amount of variety in its reflectiveness.

It still looks very flat, however, and we can't add that detail with faces since it would take thousands of them to properly display all the detail that the texture shows. This is where the normal map comes into play.

Normal map

You might remember what a normal is: the direction that a face is pointing in. This is basically the logic of a normal map.

It fakes the details by telling Blender where the supposed faces would be pointing, using a combination of colors that results in a mostly purple-looking map in most cases. If we then apply this map to our surface, we get this:

Figure 4.8 – Normal, roughness, and color maps applied to the surface

Now this looks much better and very realistic. All the little creases and imperfections are very clearly displayed.

You might have noticed that there are two versions of the normal map that have been downloaded. One is the inverse of another on the green channel; this is to cater for how different programs interpret normal maps. Blender uses the OpenGL format, so we'll use the normal map image with the suffix "GL" in the case of ambientcg, which names its textures like this. The "DX" suffix in the other normal map indicates that it was made for the DirectX format, which some programs use.

You might also have noticed that there's a **Displacement** texture map along with all the others, but in our case, we won't use it, as this map is used to physically deform a geometry according to the details in the texture. As we've said, though, this requires thousands of faces to achieve, so this is only ideal if you're working in a closeup pre-rendered animation.

Now it's time to actually apply the textures to our models.

Applying the textures

Back to our earcups' materials, in Object Mode and with our earcups selected, we should go to the **Materials** tab where we added the two blank materials, select the second material, and click the **New** button. This will add two nodes in **Shader Editor: Principled BSDF** and **Material Output**:

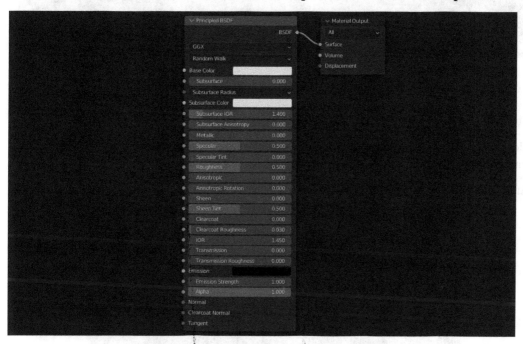

Figure 4.9 – The Principled BSDF and Material Output nodes

The **Material Output** node is what will display the texture we connect to its inputs (**Surface**, **Volume**, and **Displacement**); the **Principled BSDF** node is what controls all of the physical properties we have covered so far, and its inputs are where we will connect all of our maps.

To import our textures, we can press *Shift + A* (the same way we add another object in the 3D Viewport), look for **Image Texture** in the search bar that pops up, and then select **Image Texture**; you should see an orange node appear in **Shader Editor**:

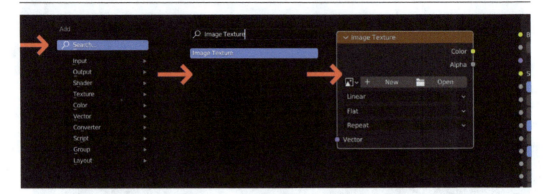

Figure 4.10 – Adding an Image Texture node

This is the method we'll use to add any node we need.

Now, to tell Blender which images to import, we click **Open** and select our image from Blender's internal file explorer. We'll start with the color.

After selecting the color texture, the node should display the image's name, along with a few more options:

Figure 4.11 – Image Texture set to the color map

Now, we can either do the same with the roughness and normal maps we will use or duplicate the node by selecting it and pressing *Shift + D*, just like in the 3D Viewport. Then, to select another image, click on the **X** button to the right in the **Image Texture** node and select another image with the same method.

An even easier method would be to simply drag each image directly from your files into **Shader Editor**, but this only works with image textures; the rest of the nodes will have to be added manually or duplicated.

After that, you should have something like this:

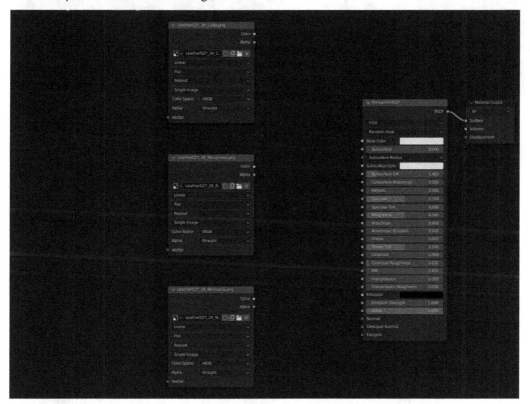

Figure 4.12 – The color, roughness, and normal maps imported

Before we connect all the textures to their respective sockets in the **Principled BSDF** node, we need to tell Blender that, for the roughness and normal maps, the colors don't actually represent colors, but rather quantitative data. To do that, we need to change the **Color Space** setting from **sRGB** to **Non-Color**, by selecting it from the drop-down menu:

Figure 4.13 – Setting the Color Space setting to Non-Color

After setting up the color space for both the roughness and normal maps, we can now start connecting the nodes to make our material finally show up. To connect one node to another, simply click and drag on the colored dots to the side of the property.

Again, this process is very intuitive: the color map goes into the **Base Color** socket, and the roughness map goes into the roughness socket (for a gloss map, you'll have to connect it to an **Invert** node and connect it to the **Principled BSDF** node).

The normal map requires one more node to work, though. We need to add a **Normal Map** node and connect our image to the **Color** input. This will convert the quantitative data from the normal map into normal data. Then, we can connect the **Normal** output from the **Normal Map** node to the **Normal** input from **Principled BSDF** (you can adjust the strength of the bump effect by adjusting the **Strength** slider). Your node setup should now look like this:

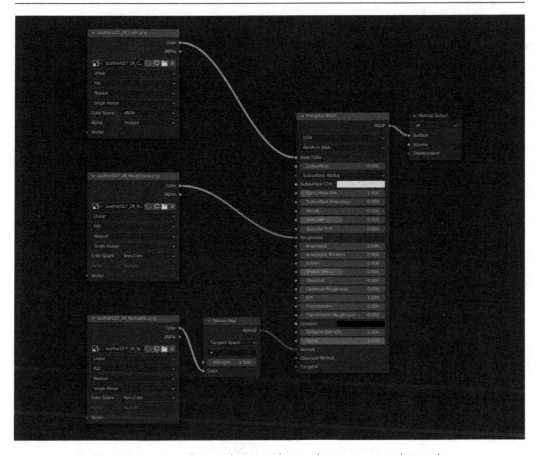

Figure 4.14 – The color, roughness, and normal maps connected properly

Now our texture should show up in the faces we assigned this material if we go into the **Material Preview** mode:

Figure 4.15 – Leather material displayed

Great! The leather itself seems right, but if we look closer, we notice that it looks a bit bigger than what you'd actually find in the real world. If we look at references, leather in the real world has much smaller creases and cracks, so we need to adjust the scale of our texture.

To do that, we need to add a **Texture Coordinate** node, which will get the coordinates of our texture and connect them to the **Vector** input of a **Mapping** node, which allows us to manipulate the position, rotation, and scale of our textures, and connect the **Vector** output of the **Mapping** node to the **Vector** input of all our image textures. We can now adjust the **Scale** sliders from our **Mapping** node. You should have a setup like this:

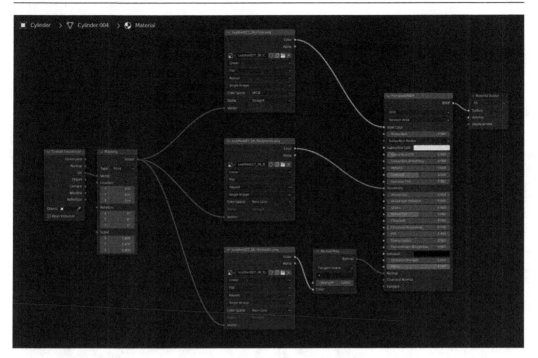

Figure 4.16 – Basic PBR material setup

For our example, a scale of 5 . 4 on all three axes was enough to make it believable. It's worth noting that all three sliders on the **Mapping** node should have the same value, otherwise the textures are going to appear stretched. We can change all the sliders at once by clicking on the topmost **Scale** slider (for the x axis), dragging down to the slider at the bottom (for the z axis), and then sliding right or left.

We can also control them by connecting a **Value** node to the **Scale** input of the **Mapping** node:

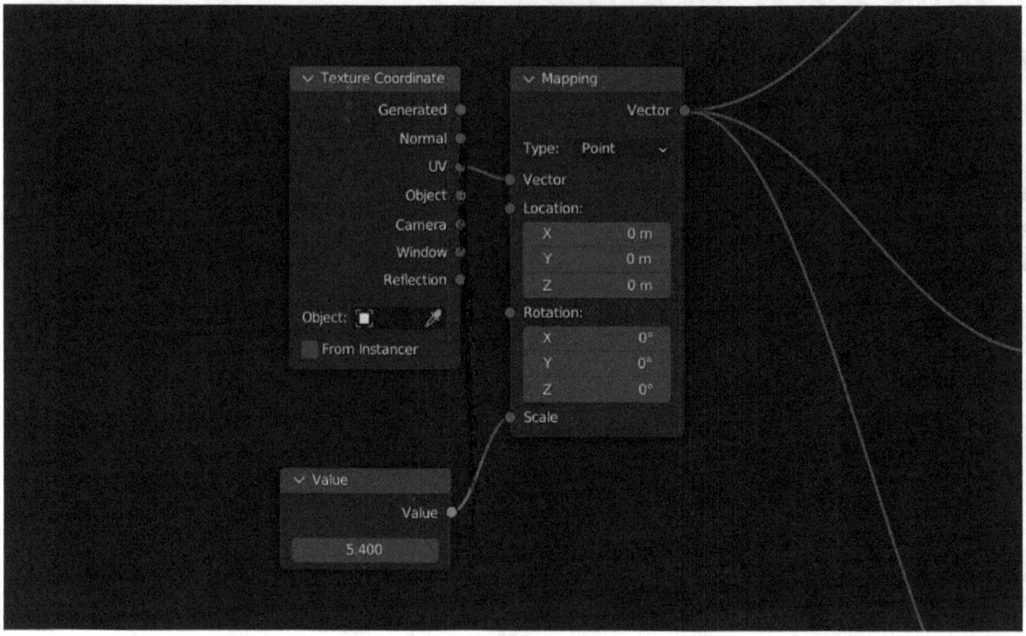

Figure 4.17 – Controlling the texture's scale using a Value node

Now, we could do that for every PBR texture we want applied to our model, but that's too time-consuming. We can instead let Blender do it for us with the assistance of the built-in **Node Wrangler** add-on.

The Node Wrangler add-on

Blender has a built-in set of add-ons that can be enabled using the **Preferences** menu. Add-ons can be very helpful in speeding up your workflow. We'll enable just one for now (though you can play around and look for anything else you might find useful).

The **Node Wrangler** add-on offers a long list of shortcuts and features to make texturing faster. To enable it, we need to click on **Edit** in the top-left corner of Blender's main window, then select **Preferences…** from the menu that comes up:

Figure 4.18 – The Edit menu

After selecting **Preferences...**, go to the **Add-ons** tab on the left (**1**), enter node wrangler in the search bar (**2**), and then enable the add-on by selecting the checkbox (**3**):

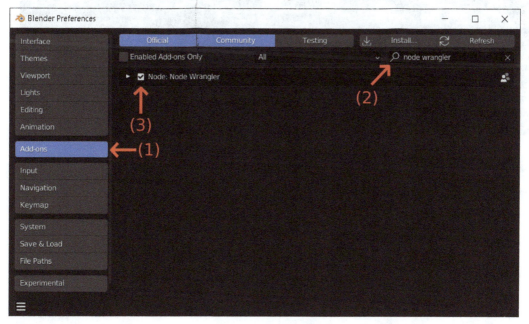

Figure 4.19 – Enabling the Node Wrangler add-on

Now, we should be able to set up PBR materials automatically in most cases. For that, do the following:

1. Select your object in Object Mode.

2. Go into **Shader Editor**.

3. Select the **Principled BSDF** node.

4. Press *Ctrl + Shift + T*.

5. Inside Blender's file explorer, select the necessary maps (holding *Ctrl* will let you select multiple maps at once, and holding *Shift* will let you select multiple maps too but will also select every file in between the first and last selections).

6. Select **Principled Texture Setup** in the bottom-right corner:

Figure 4.20 – PBR material automatic setup

You should see the same setup we had earlier for the leather, but with groups around our nodes:

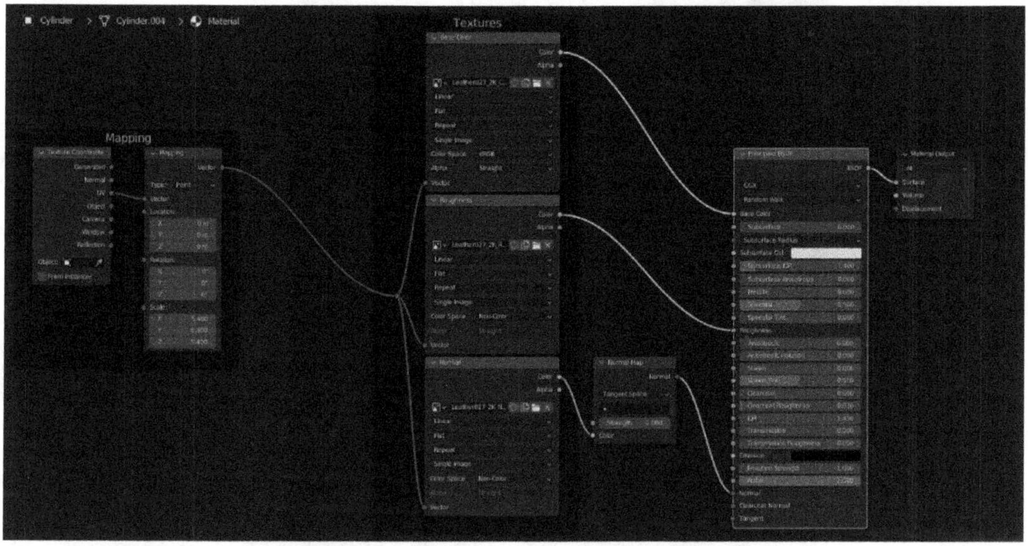

Figure 4.21 – Node Wrangler's default PBR material setup

Perfect – now we can assign PBR textures much easier than before and know where to connect each texture in case the add-on doesn't pick something up (as can happen sometimes).

But what if you want the same material for more than one object? Let's say, for example, that we want our headband's cushion to be made out of the same leather as our earcups' cushions. Well, we can just assign this same material to the other part. The process is almost the same: select the object, add two blank materials, select the faces to whichyou want the material to be applied, and click **Assign** in the **Material Properties** tab; this time, though, we won't create another material. What we'll do is select the same material we just created by selecting it in **Material Browser**, to the left of the **New** button:

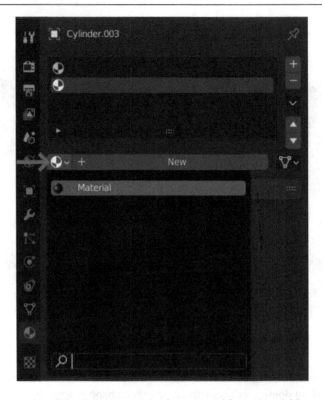

Figure 4.22 – Selecting the same leather material from Material Browser

Then, you should see the material applied in the **Material Preview** mode:

Figure 4.23 – Leather material applied to multiple objects

As you can see, the material we created was applied to the headband's cushion too. This method is great if we want to change the properties of two parts at once. We'll use the same method to texture the inside part of our cushions (add another blank material, assign it to the desired faces, click **New** in the **Material Properties** tab, and then apply the material in **Shader Editor**), which in most cases is made out of cloth to better let the sound pass through.

For that, we'll use the **Fabric 029** texture, also from `ambientcg.com`:

Figure 4.24 – The Leather 029 texture from ambientcg.com

After being applied properly, the earcups' cushions should look like this:

Figure 4.25 – Earcups' cushions, rendered with raytracing

Perfect! The cushions are looking very good, although the cloth needed to be darkened to get the right color. You can change the color, saturation, and brightness of any texture with a **Hue Saturation Value** node, which in this case was put right after the base color map:

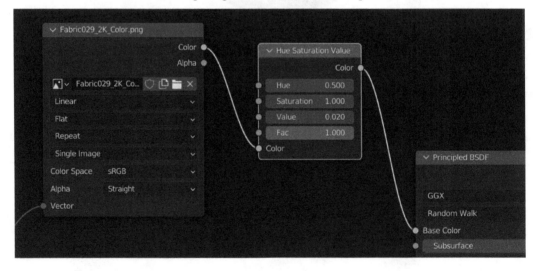

Figure 4.26 – The Hue Saturation Value node added after the color map

As you can see, the **Value** slider was lowered from the default value of 1 to 0.02, only 2% of its original brightness.

Now, for the next three textures of the earcups, we'll use another method of texturing, one that grants us more control over the aspects of the texture: procedural texturing.

Procedural texturing workflow

Procedural texturing is the term used to describe a workflow in which, instead of images being used to texture a model, nodes are used to generate and manipulate quantitative data and then turn it into color, roughness, and normal information, which usually involves doing a bit of math. The advantages of this method are that it gives the user full control of every aspect of the texture and that it has infinite resolution (until we get into the baking of the textures into images, which will be explained later on), though it takes a bit longer to display inside Blender due to the calculations being done.

Let's start with the simplest thing and increase the difficulty for each part of our earcups.

Glowing materials

Let's give our headphones some glowing areas in the earcups.

We'll start by adding a solid white glow to the loops extruded inward around the earcup's back part (marked in orange):

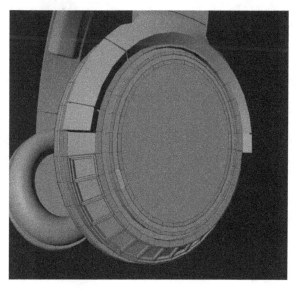

Figure 4.27 – Face loops that will glow solid white

Assign a new material to this part, hop into **Shader Editor**, and simply add an **Emission** shader connected to the **Surface** input of the **Material Output** node:

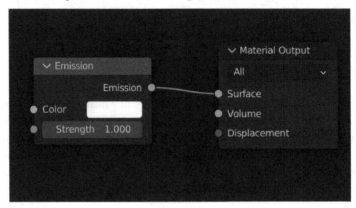

Figure 4.28 – Emission shader

Now, you can play with the strength and color to your liking. It's worth noting that less saturated colors tend to glow more intensely.

We settled for a solid white color and a strength of 5 . 6, and the result should look like this:

Figure 4.29 – White glow applied to the earcup's detail, rendered with raytracing

This can help give our headphones a more modern look. Just like that, the first procedural material is done!

When we told you we were going to start out simple, we meant it!

Now it gets a little more difficult, though, as we add more glowing materials, but this time with multiple colors on the same material, in the form of a gradient.

Multiple color glow

Now, we'll add another glowing material to the grill-like structure we extruded inward as tertiary detail (marked in orange):

Figure 4.30 – Detail we're going to add a more complex glowing material to

We'll start with the same node setup as the simple glow material, but this time we're going to use a few nodes to drive the **Emission** node's color input.

The main texture we will use is a gradient texture:

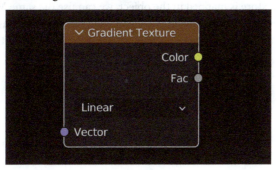

Figure 4.31 – Default Gradient Texture node

As you might expect, this texture generates a gradient in our mesh, but at the default **Linear** setting, the gradient goes from left to right or top to bottom (or the inverse of both, depending on how we manipulate it with a **Mapping** node), and what we need is a gradient that goes around a certain coordinate, like the hands of a clock when turning.

Conveniently enough, there's a **Radial** setting that does just that, so we can go ahead and change it from **Linear** to **Radial** in the drop-down menu:

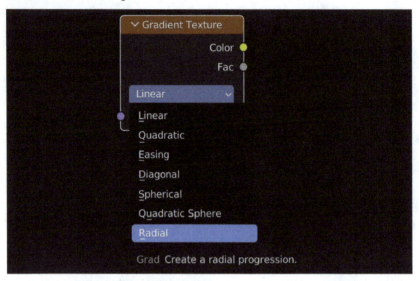

Figure 4.32 – Changing the gradient type to Radial

Now, to control how this texture is mapped in the object, we can use the same **Mapping** and **Texture Coordinate** nodes we used in the PBR workflow. If you have the **Node Wrangler** add-on enabled, you can quickly add these nodes that are already connected to the **Gradient** node's **Vector** input using *Ctrl + T*.

This is our texture setup right now:

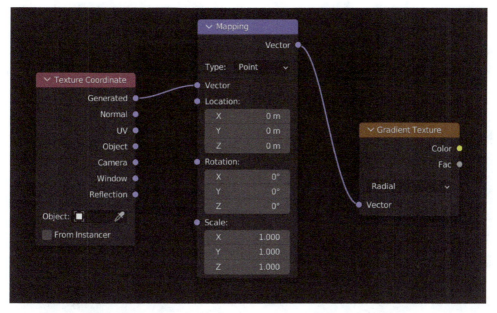

Figure 4.33 – Base node setup for the colors gradient glow texture

We can preview how our gradient looks by selecting the node we want to preview (the **Gradient** node) and using the shortcut *Ctrl + Shift* and left-clicking on the node (this only works with the **Node Wrangler** add-on). This is our current preview:

Figure 4.34 – Gradient preview

As you can see, there's not much of a gradient. That's because the texture coordinate we're using is incorrect, as the default **Generated** coordinate uses the location of the vertices to map the texture onto the surface.

A better option would be to use our earcup's origin as the texture's origin as well, since it's centered and we want our gradient to go around the center. For that, we connect the **Object** texture coordinate to the **Vector** input of the **Mapping** node, swapping the previous coordinate.

The preview should look like this:

Figure 4.35 – Gradient texture with the earcup's origin as a center

Now, we can see more of a gradient going on. We can control its rotation by using the *z* rotation slider in the mapping node. We settled for 180° of rotation since we went with the end of the gradient to get a smoother transition between the colors (you can play around with the values to your heart's content – see what looks better to you!). This is our current node setup:

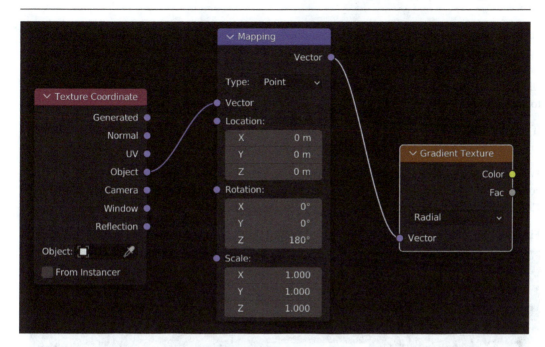

Figure 4.36 – Current node setup

Alright – now it's time to add our beloved colors, as our texture is only black-and-white right now. To do that, we'll connect our **Color** output from **Gradient Texture** to a **ColorRamp** node:

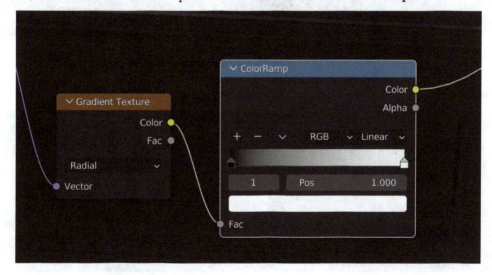

Figure 4.37 – The ColorRamp node

The **ColorRamp** node converts anything connected to it into a range of colors, defined and controlled by the little stops in the slider. As of now, the stops are set to black and white, respectively, but we can change them to the colors we want our gradient to have when glowing by selecting a stop and clicking on its respective color at the bottom part of the **ColorRamp** node – that will give you access to a color wheel.

If you're following along, you can copy our settings, pick your own colors, and/or add more colors (again, do what you think will look better; the more you play around, the more you'll find out).

We used a blueish green (HEX code: A9FFEB) to replace the black stop and a more saturated blue (HEX code: 2D38FF) to replace the white stop. Then, we connected the **Color** output of **ColorRamp** to the **Color** input of the **Emission** node, which was set to a **Strength** of 15 due to the more saturated colors.

Finally, to get an even smoother gradient, we changed the **ColorRamp** type from **Linear** to **Ease**:

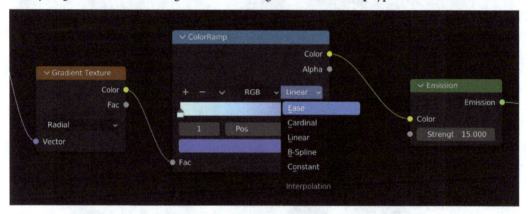

Figure 4.38 – Changing the ColorRamp type to Ease

And with that, this is our final node setup:

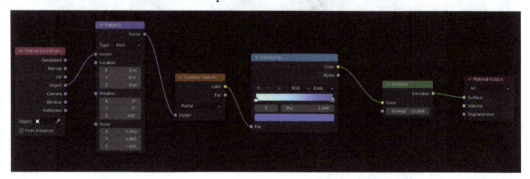

Figure 4.39 – Final node setup

You can drag the stops left or right to make a color more or less dominant and adjust the settings to fit your style. Our second glowing material is done!

This is the final result:

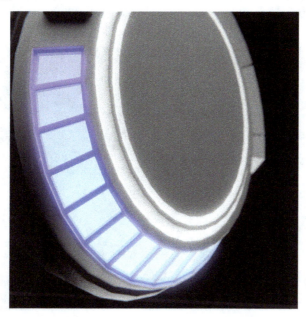

Figure 4.40 – Final result for the color gradient glow, rendered with raytracing

Perfect – this looks very good. Now we'll step up the difficulty a little and make some plastic, which is what most of our product is made of.

Plastic

This is the predominant material of our headphones, so it will be applied to most parts/faces. If we were to select the faces we wanted the material to show up for, though, it would take much longer than necessary, so we can instead select the remaining faces we *don't* want the material to be assigned to, assign a *new* blank material to those faces, and use that first blank material (which we mentioned earlier in the chapter) that gets applied to every face. On to the nodes we go!

Now, in contrast to what we did with the glowing materials, we'll actually use the default **Principled BSDF** node for the plastic.

To start, we can set the main color with the **Base Color** input from **Principled BSDF**, which lets us pick any color from a color wheel. We chose a black color, although not perfect black (HEX code: 1F1F1F) since pure colors don't appear in the real world (we've already assigned it to all the parts of the headphones that are going to have this material).

Now, we're going to work on the tiny little bumps most plastics have all over them. Although this is a very subtle detail, it can add a lot when we get closer to the model.

We can start by adding a noise texture and previewing it using *Ctrl + Shift* and left-clicking on the node, which will give us this result in texture preview mode (two of the three parts this texture is applied have been isolated for better preview):

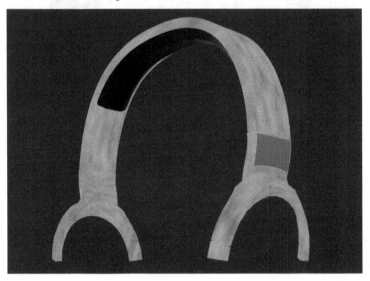

Figure 4.41 – Default noise texture preview

As we can see, there is a little bit of resemblance to the pattern we're looking for, but it lacks contrast, it's stretched, and the black spots are way too big. We're going to fix the stretching using a **Mapping** node and a **Texture Coordinate** node (which can be added automatically if you have the **Node Wrangler** add-on enabled, as explained previously).

With those nodes added, we can see that Blender is using the **Generated** coordinates to map the texture to the surface:

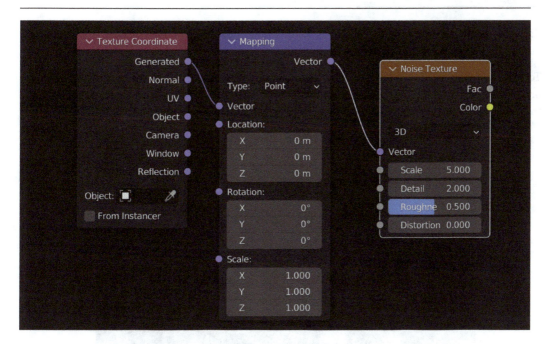

Figure 4.42 – Default coordinates used by Blender to map the texture

This is probably what's causing the stretching because it takes the position of every vertex into account when mapping and maps them in a range from 0 to 1 (sometimes this coordinate works too, so don't worry too much about why it doesn't work in this specific case – you can test and tweak multiple coordinates until you get the desired result).

We'll change that to the **Object** coordinate for this one, and immediately you will notice that the texture is not stretched anymore:

Figure 4.43 – Object coordinate used to drive the Vector input of the noise texture

Right now, the texture generated by the **Noise Texture** node looks like it's dirty, but we want to replicate bumps on plastic. We'll change some settings in the texture itself.

To start, we left the **Scale** parameter at the default 5 (you may require a different number depending on the scale of the object and which type of mapping you decide to use), then we set the **Detail** slider to 15, as right now the texture looks blurry. This is what we have now:

Figure 4.44 – Noise texture preview with updated settings

Now, the last setting to be changed is the roughness, which we'll change to 1 to make the surface look rougher, like the plastic we're going for (this can also be useful for creating a concrete material, for example). This is what we're left with:

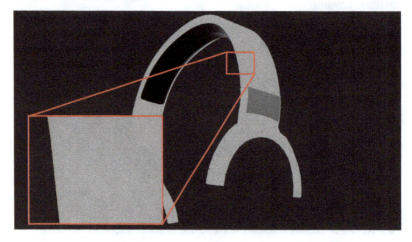

Figure 4.45 – Final noise texture settings preview

As you can see, this looks much more like the result we're going for, but it needs more contrast to appear nicely, since black values will appear to have been pulled inward, while white values will be pulled out to create the bump illusion. If we leave this as it is, the bump will be very faint. To control that, we'll use the **ColorRamp** node, with the **Factor** output from the noise texture driving the **Factor** input of **ColorRamp**:

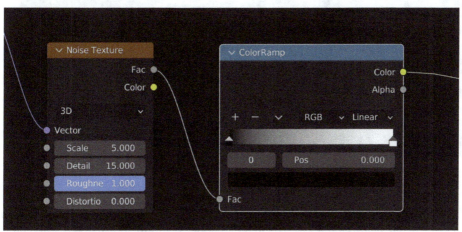

Figure 4.46 – ColorRamp added to the node setup

Now, we'll increase the contrast by sliding the black stop right and the white stop left, as this makes the transition between the white and black less smooth. We left the black spot at 0.265 and the white at 0.676. We're left with this preview:

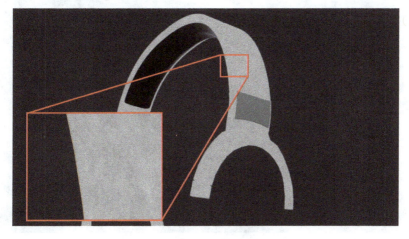

Figure 4.47 – Contrast added using the ColorRamp node

Perfect – now we can turn this texture into bump information with a **Bump** node:

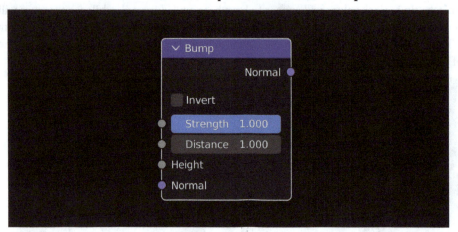

Figure 4.48 – Bump node

We'll connect the **Color** output from **ColorRamp** to the **Height** input of the **Bump** node (since it's a black-and-white map), connect the **Normal** output of the **Bump** node to the **Normal** input of the **Principled BSDF** shader, and connect its **Shader** output to the **Surface** input of the **Material Output** node:

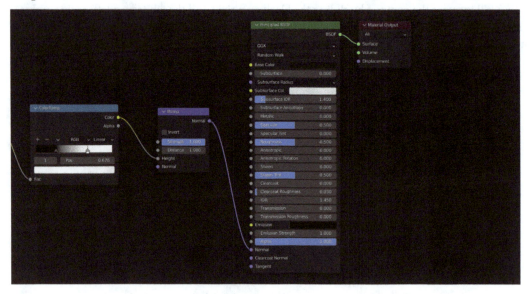

Figure 4.49 – Bump setup

After making these connections, you should see this in **Material Preview**:

Figure 4.50 – Bump preview

As you can see, this is too strong, but it worked nonetheless. What's left for us is to play with the **Strength** and **Distance** settings from the **Bump** node until it looks a lot more subtle. We ended up with a strength of 0.063 and a distance of 0.002. This should be the result:

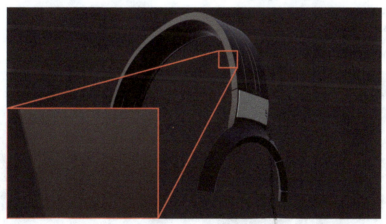

Figure 4.51 – Final bump preview

As you can see, the bump is very subtle, but it helps break that perfection that computers love to leave us with by default.

Another thing we can do to help break that perfection is vary the roughness of the surface a little bit, as nothing in the real world has a constant shininess across the entire surface; this will also look subtle.

We'll use a node called **Musgrave Texture**, which looks like this:

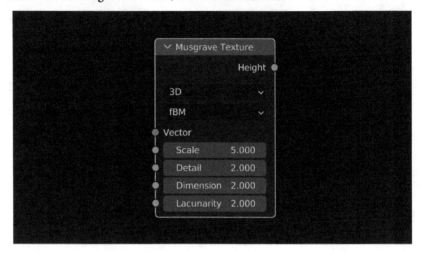

Figure 4.52 – Musgrave Texture

To start, we'll connect the same mapping node we used for the noise texture to the **Vector** input of **Musgrave Texture**, as if we ever want to change the scale or rotation of the texture, everything will change by the same value.

If we preview **Musgrave Texture** now, we should see this:

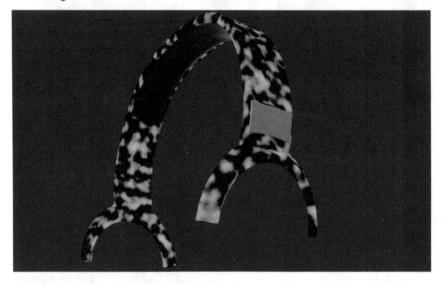

Figure 4.53 – Preview of the default Musgrave texture

This also looks too blurry, so we need to tweak some settings as well: first off, **Scale** was kept at 5, **Detail** was set to 15, like the noise texture, and **Dimension** was set to 0.8, to make it even more detailed and to have the detail appear more. The lacunarity was kept at the default. This is what you should see now:

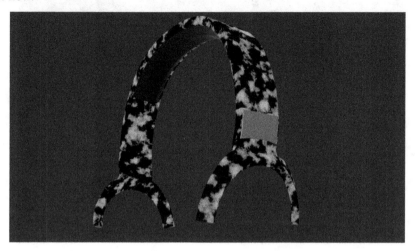

Figure 4.54 – Final settings for the Musgrave texture previewed

Again, it looks like we're making the headphones dirty, but we're really making them just a bit less perfect by adding a subtle amount of variety to the roughness of the surface.

As of now, though, if you plug the texture into the **Roughness** input of the **Principled BSDF** node, you'll see a very extreme effect because of the contrast between the black and white – how do we reduce it? With **ColorRamp**, of course!

To decrease the contrast of a texture with a **ColorRamp** node, we need to connect the Musgrave texture to **ColorRamp**, and then change the colors of the stops to more similar colors. We'll keep it in grayscale in order for it to work as a roughness map, but we'll make both stops have similar tones of gray (HEX code for the left stop: 5A5A5A; HEX code for the right stop: 7A7A7A), which will look like this when previewing:

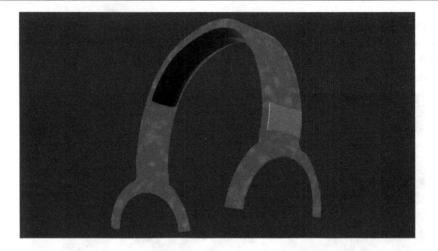

Figure 4.55 – Decreased contrast on the texture

Now, if you want to control how shiny it is, you can play around with the colors of the **ColorRamp** stops, but we can change it all at once without affecting the contrast we chose by using a **Math** node, which does, as the name suggests, math operations with the values provided:

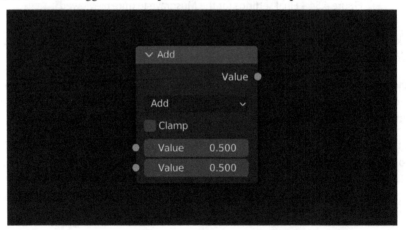

Figure 4.56 – Math node

This node has a variety of uses and operations (which can be seen by selecting the menu in the upper part of the node), but we'll stick with the **Add** mode.

To use the node, we connect the output from our **ColorRamp** to the **Value** socket from the top of the **Math** node; then, if we preview this node, we see this in our mesh:

Figure 4.57 – Math node preview

As you can see, the texture turned almost entirely white, and if we connect the **Math** node to the **Roughness** input from the **Principled BSDF** shader, we'll see little to no reflection because of the **Math** node. We'll lower the bottom value to add less roughness, therefore making it shinier in this case (if you reached 0 and are still not satisfied, negative values are possible – or you can change the mode to **Subtract** using positive values). We kept it at a value of 0.250.

This is our final node setup for this material:

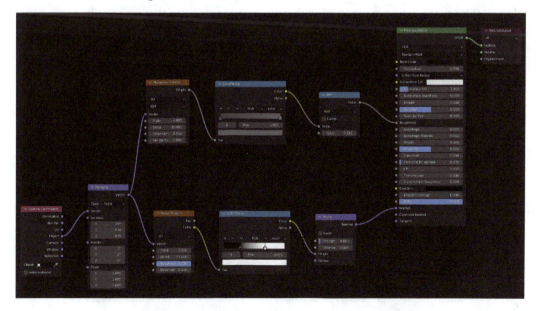

Figure 4.58 – Final plastic node setup

If we then connect the **Principled BSDF** shader to the **Surface** input of the **Material Output** node, we should see the finished material:

Figure 4.59 – Finished plastic material, rendered with raytracing

Great – this is starting to take shape. Now we'll texture the metallic part we usually see on headphones.

Metallic part

Looking at this reference, we can see an effect that usually happens when there's a metallic surface on the back of the earcups of a pair of headphones:

Figure 4.60 – Metallic material on headphones (image from https://www.pexels.com/)

This effect is called anisotropy and happens in brushed metals – it can also be seen in pans, for example.

Although the **Principled BSDF** shader has a setting to control how anisotropic a material actually is, some render engines don't have that feature, so we're going to have to replicate that effect by manually adding the bumps caused by the brushing using procedural textures.

To start, we'll need to understand how this effect happens in the real world: upon brushing the metal, tiny bumps are left, which reflect the light in a specific way, causing this effect. Those bumps are sometimes purposefully laid out as concentric circles to make the effect look better. We'll recreate that.

To start, we'll establish where the center of our circles actually is by using the object's physical coordinates. To do that, we'll need the same **Texture Coordinate** node connected to a **Mapping** node. Instead of connecting this to a texture, however, we're calculating the distance from the center of those coordinates.

We'll use the **Object** coordinates from the **Texture Coordinates** node to drive the **Mapping** node, then connect that to a vector math node, which allows us to do a variety of math operations to manipulate the coordinates:

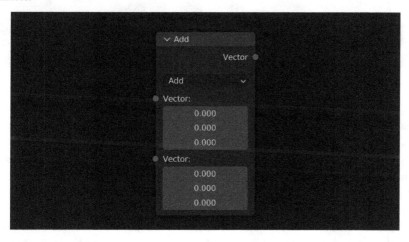

Figure 4.61 – Default vector math node

We'll not use the default **Add** mode, though, as this only manipulates the coordinates in 3D space, and we can already do that with the **Mapping** node. What we will use is the **Length** mode, which can be selected from the drop-down menu:

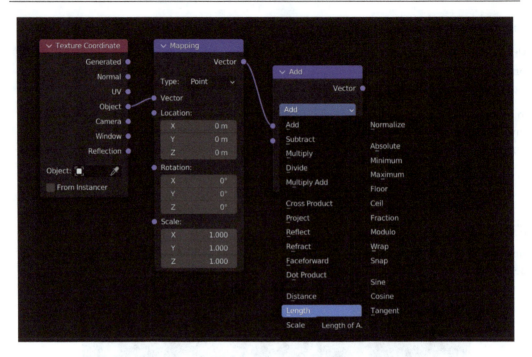

Figure 4.62 – Length mode of the vector math node

This mode has no settings because the only purpose of it is to calculate the length from the origin of the texture in all directions, which will result in a perfect circle. The origin is a black dot (with a value of 0), which you won't see if you preview the node. That's because we're using the object's coordinates to calculate this, so the black dot is at the origin of our object and is therefore not visible on the surface. We can manipulate this with the sliders in the **Location** parameter from the **Mapping** node, so we'll manipulate it until we get that black dot in the center of the earcup's back (you can make the sliding more precise by holding *Shift* while changing the slider). Those specific values may vary from object to object, even if you're following along, but for our headphones, we left the **X** coordinate at 0 m, the **Y** coordinate at -0.1 m, and the **Z** coordinate at 0.45 m. You should see this when previewing the **Length** node after the coordinate manipulation:

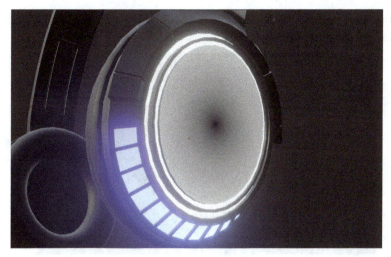

Figure 4.63 – Length node manipulated to the surface

Alright, we have a circular coordinate going on – now we need to make those concentric circles that almost look like waves… Wouldn't it be nice to have a wave texture?

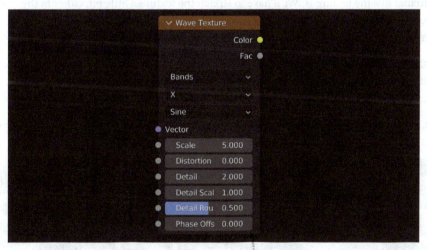

Figure 4.64 – Default Wave Texture

This texture generates a wavy pattern, alternating from black (0) to white (1); the only thing is that, in its default state, it makes straight waves, and we need circular ones. That's why we made a circular coordinate!

We'll connect the **Value** output from the vector math node to the **Vector** input of **Wave Texture**; then, if we preview the wave texture, we should see this:

Figure 4.65 – Pattern generated by the current texture setup

We're getting somewhere, but the pattern generated doesn't quite fit in our headphones since it is not perfectly circular like the texture is. We'll have to stretch the texture a little bit, either in the *x* or *y* axis (in our case). We changed the **Scale** parameter in the **Mapping** node to 1.220 for the **X** slider so it fitted better with the shape of the headphones, but your values might be different.

Currently, the waves are smooth, as we can see by the fading of the edges, but we want them to be sharp, as they would be if we looked closely at a pair of headphones that had an anisotropic metallic surface. To do that, we'll round up the values to either 1 (white) or 0 (black) in order to make them sharp.

To do that, we'll connect the **Wave Texture** to a **Math** node set to **Greater Than** mode from the drop-down menu:

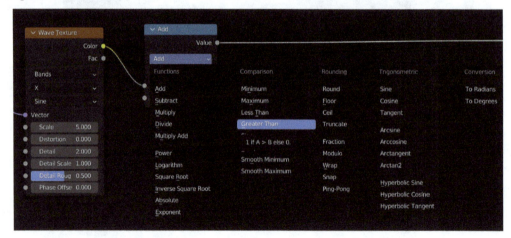

Figure 4.66 – Greater Than mode

After selecting **Greater Than** mode, you should see that the circles have sharp edges when you preview the **Greater Than** node, and you can adjust the bottom value to make the lines thicker or thinner (keep in mind that the white areas are what's going to appear to be protruding outward). We kept it at a value of 0.150, so the white areas are thicker:

Figure 4.67 – Thickened white areas

Now, the last thing left to do is to adjust how many circles we want, as this little won't make the effect show up as we wanted. To increase the number of circles, we can change the **Scale** slider in the Musgrave texture, increasing it to add more circles. We left it at 20 (keep in mind the size of the UV island in the object's UV map, because procedural textures have infinite resolution, but the final baked textures won't, so some of the detail might be hidden if it's too small).

To apply this as our bump map, we can connect the output of the **Math** node to the **Height** input of a **Bump** texture, connect that to the **Normal** input of a **Principled BSDF** shader (set the **Metallic** slider to 1 and lower the roughness to get a metallic effect), and preview the shader. This is our final node setup:

Figure 4.68 – Final anisotropic metal shader

Perfect – now let's see how this texture looks:

Figure 4.69 – Final anisotropic material

Beautiful! The effect is very nicely displayed, and with that, we're done with this part, which means that we're done with procedural texturing for this model.

It's worth noting that it is impossible to cover everything about procedural texturing, and this isn't even scratching the surface of what procedural texturing is capable of, but we covered a few of the most used setups, settings, and nodes used in this workflow.

Summary

In this chapter, we covered two different texturing methods used to apply textures to the model all at once, either by using premade textures with all of the necessary maps or by generating them using math.

Now, all of those methods are for applying premade patterns, applying textures, or using math to generate textures, but what if we want absolute control over how our texture looks? Well, then we'll need to paint them manually using texture painting, which will be covered in the next chapter, along with another slightly trickier, less common method.

Texture Painting and Using Real-Life Images as Textures

Texture painting consists of making your own textures and manually painting them onto the model itself. This process is indeed more time consuming than the methods presented in the previous chapter, but it's the ideal method in some cases.

In this chapter, we'll cover the following topics:

- How to set up the texture painting workspace
- How to paint textures manually
- Different brush settings for painting
- How to texture our models using real-life images
- How to bake textures into images in order to use them in most 3D render engines

By the end of this chapter, you should be familiar with the texture painting workflow, using and manipulating real-life images to texture your models, and baking textures into a single texture set for increased rendering performance.

Setting up your texture painting workspace

Before we start working, let's set everything up so we know what we're working with. To start, we can enter **Texture Paint** mode by selecting the object we want to paint, then either selecting the **Texture Paint** workspace option at the top or replacing **Object Mode** using the menu in the top-left corner:

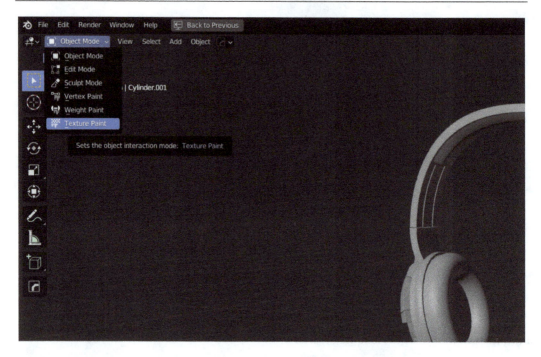

Figure 5.1 – Selecting Texture Paint mode in the viewport

Once we're in **Texture Paint** mode, we need to go to the material of the part we want to paint our textures on and add an image texture to the material. This time, instead of using an image we have already saved by clicking **Open**, we'll click the **New** option. Then, we'll see the following menu:

Figure 5.2 – Creating a new image texture

This is where we set a few parameters for our image. We'll only change the resolution and color and uncheck the **Alpha** option as this texture won't have transparency. (Make sure to give it a name as well, by typing the new name in the **Name** field.)

The **Color** attribute should be set as the predominant color of your texture and the resolution should be set at an equal or higher resolution than you want your final textures to be. We'll leave it as a dark blue color for now (hex code: 080A35) at a 2K resolution (2,048 x 2,048 px). After changing the desired settings, your image should be created and look like any other:

Figure 5.3 – Image texture set for texture painting

Now, you should be able to connect this image to the **Base Color** input of a **Principled BSDF** shader. We're ready to start painting!

First, though, let's go through some of the settings inside **Texture Paint** mode.

The Texture Paint workspace

Upon entering **Texture Paint** mode, you should see significant changes to the interface, which presents us with several settings for our brush. We'll cover the most important ones.

Here is the full viewport interface:

Figure 5.4 – Texture painting setup

We'll start from the column on the left, then work from left to right going through the top settings.

In the column on the left, we have, from top to bottom, the **Draw**, **Soften**, **Smear**, **Clone**, **Fill**, and **Mask** modes for our brush. These are mostly self-explanatory, but if you've never used a drawing program before, here are brief explanations of what they do:

- **Draw**: This is what you'll use to paint freely in your object
- **Soften**: This blurs the edges of what you painted in order to make them softer
- **Smear**: This "moves" the paint as if you were running your finger over a canvas while still wet, similar to the *Smudge* tool present in most drawing programs
- **Clone**: This duplicates a set part of the painted texture to wherever part you're currently painting
- **Fill**: This fills areas with a set color
- **Mask**: This prevents your strokes from going inside or outside a set area

Then, in the top row, we have the name of the brush we're using, as well as two colors; the left one is the one being used and it can be switched with the second one by using the *X* shortcut. Then, we have the paint mode, set to **Mix** by default, but you can paint in a variety of other modes as well to achieve different effects (which won't be covered here, but we encourage you to experiment and see how they look):

Figure 5.5 – Different modes for painting

To its right, we have the **Radius** setting, which can be set by dragging the slider, by using the shortcut *F* and dragging the cursor left or right, or even controlled by pen pressure if you're using a drawing tablet that has a pen (for that, check the pen icon to the right of the slider).

Next, we have the **Strength** slider, which controls the intensity of the applied paint. This can also be controlled by pen pressure and activated by clicking on the icon to the right of the slider. You can also adjust it using the *Shift + F* shortcut.

Then, we have the **Brush** settings. Surprisingly, there are only two:

Figure 5.6 – Brush settings for texture painting

Affect Alpha will only make a difference when your texture has transparency, as it controls whether the strokes will preserve said transparency or not. **Accumulate** determines whether the paint appears stronger when strokes overlap.

Next, there's **Texture**, which is useful when you want to manually paint a pre-made texture onto your model using it as a custom brush. Upon selecting it, we're presented with the following menu:

Figure 5.7 – Texture menu

The **Mapping** option inside the **Texture** menu brings up another menu with options for how our brush is applied to the surface. We'll get more into that later.

The **Texture Mask** option has the same menu, but it works with masks. We won't use this option for this model, though.

The **Stroke** menu, however, is very important and controls how the brush strokes themselves will behave when applied:

Figure 5.8 – Stroke menu

The most important settings inside this menu are **Stroke Method**, which controls how the strokes are made, **Spacing**, which controls how spaced out the brush's texture is in a stroke, **Jitter**, which controls how random the rotation is in each stroke (can be useful for creating organic textures), and **Stabilize Stroke**, which smooths out the stroke and can be very handy if you're not working with a drawing tablet.

Last, but certainly not least, we have the **Falloff** option, which controls how faded the brush's texture is inside the brush stroke (radially).

Perfect, now we can finally paint our model.

Painting

Now that we know what tools we're working with, we can effectively paint some good-looking textures. We'll show you three uses for texture painting as well, using those same tools. Remember: functional UVs are essential for this to work.

Manual color painting

This is the default painting method inside Blender. The brush will appear around your cursor, as a circle. You can adjust the scale and intensity of it by using the *F* shortcut and then dragging your cursor left or right and *Shift + F* and dragging your cursor, respectively.

To actually paint, you can just select the image you created in the materials, then in **Texture Paint** mode, hold the left button and drag your cursor across the surface to create any drawing, pattern, or texture you want. Here are two examples:

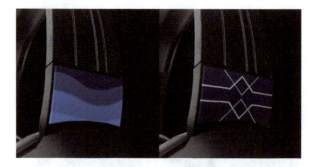

Figure 5.9 – Examples for texture painting

The left example was painted freehand, while the other was made using the **Line** mode instead of the default **Space** mode for **Stroke Method** (there are various modes; you can test them to see which looks better). This can be set inside the **Stroke** menu, as seen in the following screenshot:

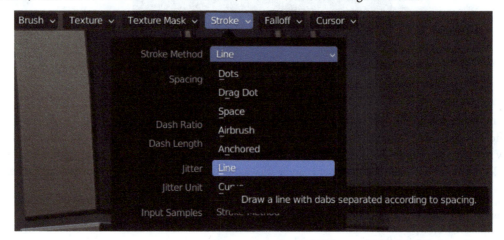

Figure 5.10 – Setting Stroke Method to Line

Upon selecting this mode, you can click and drag to make the line, which will be applied after you release the left mouse button.

You could paint your entire model with this workflow, and you can also use **Roughness**, **Metalness**, **Bumps**, and other maps if you use only grayscale colors and connect the image textures accordingly.

Let's see another use for texture painting.

Mixing textures

Besides painting colors manually, you can also use grayscale colors to mix pre-existing sets of textures.

Let's say, for example, that you want to add some sort of pattern with color variation in the same texture and you want to have total control of the placement of the variations. You can use texture painting for that. We'll demonstrate it using a camo pattern applied to the same area as the previous examples.

Let's say you have two PBR texture setups, one for a lighter camo and one for a darker camo:

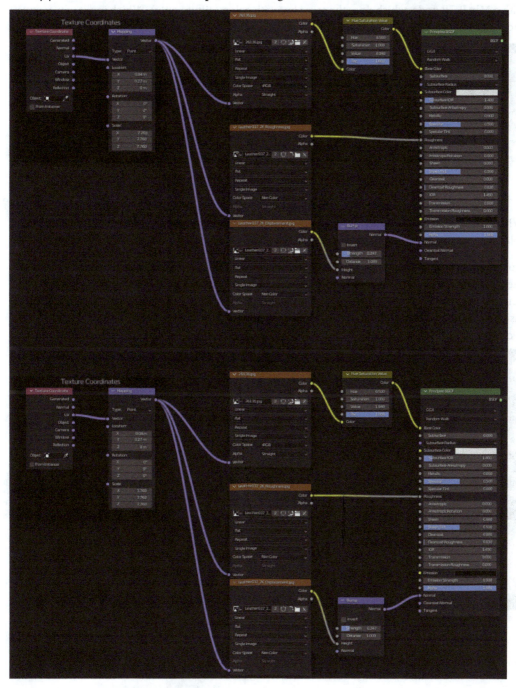

Figure 5.11 – Two PBR node setups for lighter and darker versions of the same texture

To mix the two, we'll connect both **Principled BSDF** nodes to a **Mix Shader** node (not to be mistaken with a **Mix RGB** node, which does the same but with colors/diffuse only). Which input you connect each of them to won't matter much in this specific case:

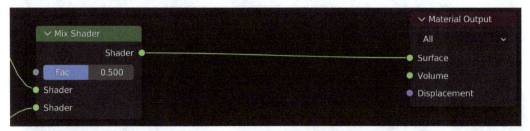

Figure 5.12 – Principled BSDF nodes connected to a Mix Shader node

What we connect to the **Fac** input determines which textures appear, using values that range from 0 (black/upper slot) to 1 (white/lower slot).

Now, we do the same thing with the image textures: add an **Image Texture** node, create a new image, then connect the **Color** output of **Image Texture** to the **Fac** input of the **Mix Shader** node and start painting in grayscale only. Remember that anything you connect to the top **Shader** input will be more visible when the value is closer to 0 (black) and the bottom input will be more visible when the value is closer to 1 (white):

Figure 5.13 – Image texture connected as the Fac on the Mix Shader node

Now, we can paint some stripes of lighter camo on top of the darker camo, to make our texture interesting:

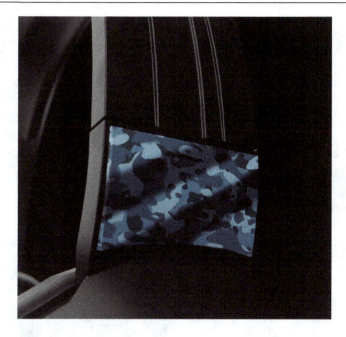

Figure 5.14 – Two sets of PBR textures mixed using texture painting

This technique can be very handy if you want, for example, to add moss to a wall or a pillar, or maybe add a snow layer to a ceiling or a car.

Now, let's have a look at one more tool/use for texture painting.

Stencils

Stencils in Blender work just like they do in real life, except here, you'll need an image, either with transparency or not, to apply the stencils to the object properly. In our case, we'll use a logo with a transparent background. We'll apply the logo stencil in two places on our headphones, the camo leather we just textured and the metallic back of the earcups, since it's common to have a logo in those areas, which you'll probably notice if you look at some references.

In order to add a stencil, we'll have to add either another image texture for direct application on the surface or another texture setup. In our case, we'll do something similar to what we did for the previous texture, by adding yet another **Principled BSDF** shader to our material, and use the result of this painting to determine where our logo will appear using another **Mix Shader** node (by using this method, we'll have control over the color and texture of our logo, if we want to change any of it afterward):

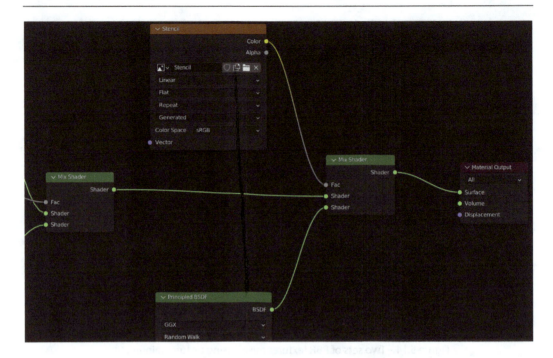

Figure 5.15 – Mixing the camo material with another Principled BSDF shader
for the stencil, using an image texture as the driving factor

To get the stencil working, we'll head over to the **Active Tool** menu on the right while in **Texture Paint** mode with our brush:

Figure 5.16 – Active Tool menu

Now, inside that menu, we should scroll down until we find the **Texture** menu:

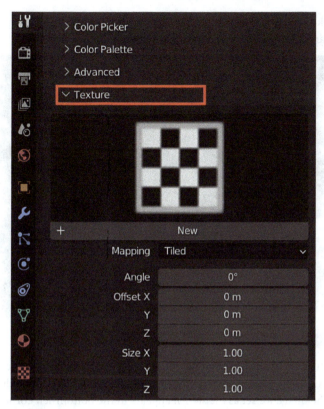

Figure 5.17 – Texture menu

From there, we click on **New** to add a new texture. Then, to actually set what texture our stencil will have, go to yet another menu: the **Texture Properties** menu. This is the last menu in the options to the right and contains all the options we need for any textures created inside Blender:

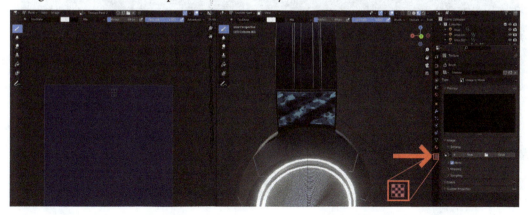

Figure 5.18 – Texture Properties menu

From there, we'll click **Open** and select our image texture from the Blender file explorer. We'll select our logo with transparency, which should appear above those same options:

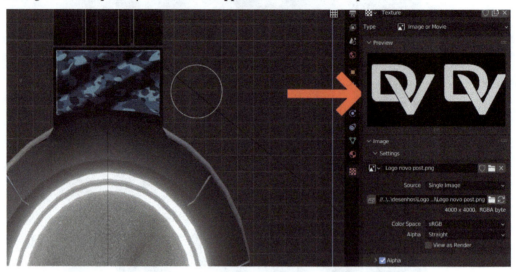

Figure 5.19 – Stencil texture set as the logo

Now, we have our texture applied as a brush. But to make it a stencil, we need to select the stencil mode from the **Texture** menu at the top of the screen, which was shown earlier. Upon selecting this mode, you should see that the image we set as the stencil appears in the bottom-left corner of the viewport window if we're in **Texture Paint** mode:

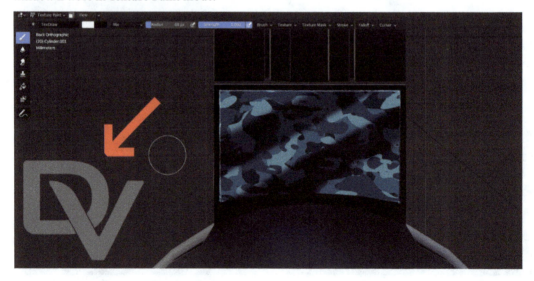

Figure 5.20 – Stencil image showing up in the viewport

To place our stencil over the area we want, we use the right mouse button to drag it around, and the *Shift* + right mouse button and *Ctrl* + right mouse button shortcuts to scale and rotate the stencil, respectively.

Now, after we place our stencil where we want it to show up, we can just paint over it without worrying about making incorrect strokes just like with real-life stencils. Also, like real-life stencils, you need to make sure you cover everything evenly, as it's very hard to get the stencil back to the position it was first applied once we move our view.

Remember, the stencil is applied based on your viewing angle, so if you apply it from any view that's not straight on, it will look distorted. For this, the orthographic view is very handy (use shortcuts *1*, *3*, and *7* on the numpad).

After applying our logo, we'll change the color to a dark cyan tone (hex code: 19262C), then we should see this as a final result:

Figure 5.21 – Final result for the stencil

Now, we'll apply this exact same technique for the metallic back part of the earcups. We encourage you to try to apply this on your own and experiment with other colors, patterns, images, and so on. It's all about playing around.

We chose to apply a metal texture rather than just paint like we did with the leather, which ended up looking like this:

Figure 5.22 – Logo stencil applied to the anisotropic material

Perfect. Now go to the **Active Tool** menu and click on **Save all images**, since Blender doesn't store any texture-painted maps unless we tell it to do so.

Now we're ready to take a look at a simple yet very effective way of texturing any type of asset: using real-life images.

Texturing with real-life images

This method is mostly pretty straightforward, since it consists of basically taking images from the real world, strategically using all or part of them to achieve more realistic results, and manipulating said image's colors to try and make the other necessary maps.

Since our headphones are already fully textured, we'll demonstrate this method by making a corner of an outdoor wall with functional UVs. This is how it looks right now, without textures:

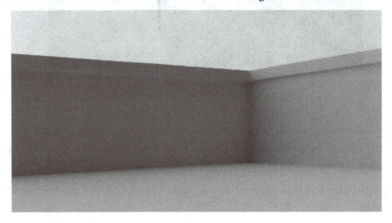

Figure 5.23 – Wall corner with no textures

To start, you should select images with high resolution, preferably even more resolution than what you want in your final asset. You can find images on the internet, on sites such as textures.com.

Ideally, those images should have flat, soft lighting instead of taken in direct sunlight, for example, which creates sharp shadows, making it very hard to relight the image with lights coming from a different direction. See the following comparison:

Figure 5.24 – Soft versus harsh lighting comparison (images from https://www.pexels.com/)

Now, with our wall texture in hand, we can start with the same usual texture setup, but instead of using a regular **Texture Coordinate** node to drive the mapping, we'll use a second UV map, which we only have to use because most images won't have square aspect ratios (which is necessary for UVs). We'll also need to break our current UV map in order to manipulate the placement of each face in the correct location. The second UV map will allow us to break our UVs and transfer the textures from the bad UV map to the good one later on when baking the textures.

To add a second UV map, we need to access the **Object Data Properties** menu, then inside it, access the **UV Maps** menu:

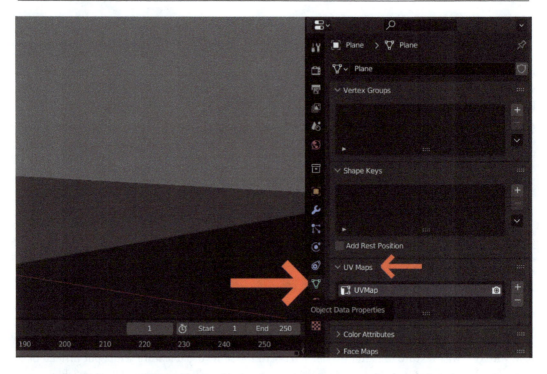

Figure 5.25 – Accessing the UV Maps menu

Now, we'll create a second UV map by clicking the plus icon to the right of the **UV Maps** menu and selecting it from that same list of UV maps:

Figure 5.26 – New UV map selected

Now, back to the shader editor; in the wall material, you should connect the desired image texture to the **Color** input of the **Principled BSDF** node, and for the **Vector** input of the **Image Texture** node, we'll use the **UV** output from a **UV Map** node, for which we'll set our new UV map as the source of data:

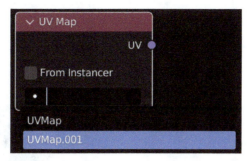

Figure 5.27 – UV Map node

Now, with everything connected, we're ready to start messing around with the UV map.

In the UV editor with the new UV selected, once the texture is applied, you should see the same working UVs from the first unwrap. Don't worry, that UV map is just a copy, which we can manipulate freely since it won't affect the functional UVs we already have.

In edit mode, we can now unwrap individual faces and/or portions of the mesh to position them in a certain part of the image and change their scale, rotation, and location to our heart's content. Don't worry about getting good UVs, as this is just where we're getting the texture data from, so you can overlap them, make them crooked, use different projections for the UVs, unwrap them based on the angle you're viewing it… You know, manipulate the UVs however you deem necessary to make them look good. Test however many options you require.

For example, this was our result for the wall's second UV map:

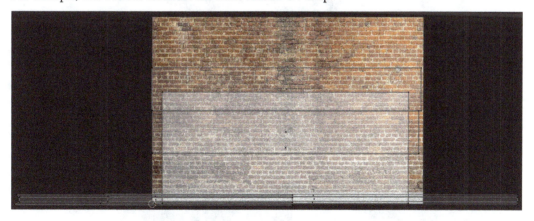

Figure 5.28 – Second UV map for the wall

As you can see, this is a mess: it has overlapping UVs everywhere, the islands go over the border of the image, and the image itself doesn't have a 1:1 aspect ratio. This is only for getting the textures, though, so it's completely fine. After positioning the UVs, you can go back to the **UV Maps** menu and select the main map again. Then, remove the image you used from the UV editor by clicking the **X** button on the top-middle part of the UV editor, so the main UV returns to the 1:1 aspect ratio:

Figure 5.29 – Removing the brick wall image from the main UV map

Here is the main UV map:

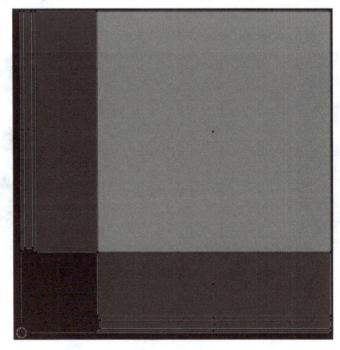

Figure 5.30 – Main UV map for the wall and floor

When we bake our textures, the texture data will be transferred to this UV map. (Baking is the process of applying all the combined textures and effects and turning the result into separate texture maps, in the form of image files, to make one single texture. This will be covered soon.)

We're still missing the other two main maps, though.

As of now, the textures on the wall look flat, to say the least:

Figure 5.31 – Wall with color only applied

To bump up the realism a bit more, we'll generate the other maps by manipulating the colors of the images we used.

If you remember, grayscale images can be used as roughness and normal maps, so we'll convert our brick wall image into grayscale using a **ColorRamp** node and tweak it according to our needs. In our case, we need most of the image to be closer to white for the roughness, since brick walls are usually not very reflective. We can preview the **ColorRamp** node to better see the results. These were our final settings and the respective preview for the roughness map:

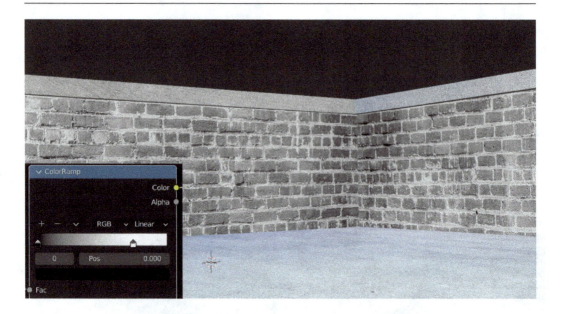

Figure 5.32 – Roughness preview and ColorRamp settings

Upon connecting **ColorRamp** to the roughness input of the **Principled BSDF** node, you shouldn't see many changes. That's because this is supposed to be subtle, just to give some variation to what was once a perfectly uniform surface.

Now, we'll give it some bumps using the same technique, but this time, keep in mind that every black spot will appear to be pulled inward and every white spot outward. Everything in between will have those effects but they will be a bit less apparent, and that's what grants our texture some more variety, giving our model more realism. According to this logic, the bricks should be lighter and the spaces in between the bricks darker. We'll achieve that through one more **ColorRamp**. Here are the settings and a preview of **ColorRamp**:

Figure 5.33 – ColorRamp settings and preview of the bumps

Now, if you apply the roughness and bump maps properly, you should get a result like this:

Figure 5.34 – Wall texture with all maps generated

If you want, you can increase the realism even more by placing edge loops around a few bricks and extruding them. This is extremely handy for something such as a building, windows, or facades. This will increase the face count, so keep that in mind along with the actual placement of these objects in the scene/world, which could determine whether you should add this type of detail or not.

With this technique applied as well, here's the final result:

Figure 5.35 – Wall with every technique applied

For comparison, here's the wall with all the techniques applied alongside only the color:

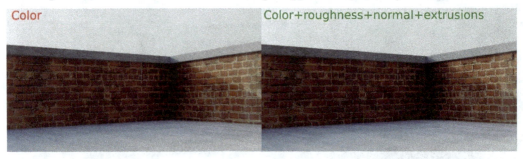

Figure 5.36 – Comparing before and after applying the techniques to increase realism

Normally, this workflow is used for objects that are not going to be the main focus but are rather elements to ensure a scene doesn't feel empty. However, you could try and use it on closer objects, depending on the quality of the image and the results.

Now, back to our headphones, let's see how to apply all those cool textures we made in one single texture set, with all the necessary maps.

Baking the textures

The process of baking the textures is essential to make our textures work outside of Blender, given that every render engine has its own method of interpreting, mixing, and displaying them. But what most of them do have in common is the use of the color (or diffuse), roughness (or glossy), and normal texture maps. So, we'll take all the tweaks, mixes, effects, and hand-painted textures we made in Blender and turn them into usable texture maps. Unfortunately, though, you'll have to either apply or delete any modifiers you may have left in order for it to work properly, so make sure you're absolutely satisfied with how your object has turned out so far.

We'll cover two baking methods: first, from high poly to high poly (meaning the exact same object, with the exact same UVs), and second, from high poly to low poly (meaning we'll transfer all of the detail from the high-poly model we've been working on in the past chapters to the one we made at the end of Chapter 2).

Remember: Working, non-overlapping UVs are essential for baking, as the final result will depend on them.

Baking from high poly to high poly

To start, we'll need to change the render engine from the default Eevee to Cycles, since it's currently the only render engine in Blender that supports texture baking. We won't cover the different features of the render engines themselves, since we're focused on modeling instead of actual rendering.

To change from Eevee to Cycles, we need to go to the **Render Properties** menu and select the **Cycles** engine from the **Render Engine** drop-down menu. Then, what's also recommended is to change the device method from the default **CPU** to **GPU Compute**, in order to increase the baking performance (only if your computer has a GPU, of course). You can see the process here:

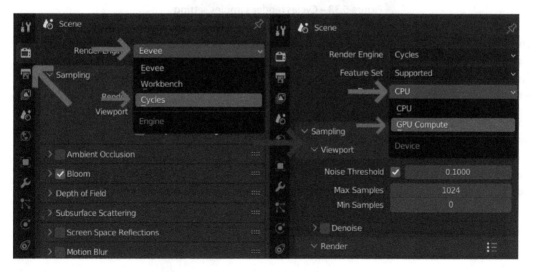

Figure 5.37 – Activating the Cycles render engine

To improve the performance even more, we can scroll down and lower the **Max Samples** value in the **Render** menu, since the baking works by taking a series of samples from our textures, with each iteration making a cleaner image.

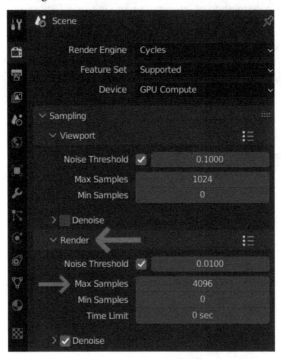

Figure 5.38 – Cycles render samples setting

The default value of 4096 is way too high for our purposes, so we'll lower it considerably, to around 240. This is so we don't make our computer compute excess data that won't make a difference in the final result. 240 samples should be more than enough to make good-looking bakes for most models, while also keeping the performance high.

To actually bake the textures of all of our objects, we need to add another image texture to all the materials of all of our objects, and create another image (remember to keep this node as the active node by selecting it last, in each material, in each object). This time, though, we set the resolution we want our final texture maps to be, and rename it to whatever map we want to bake first; we'll go with Color. Don't connect this image to anything. This is how we set up our color map:

Figure 5.39 – Color map set for baking

Now you need copy and paste this same image texture to all of your materials, across all of your objects (you can select the image you created by clicking the picture icon to the left of the **New** button in the **Image Texture** node, then selecting your image from the drop-down menu).

Now, with this setup done, we can scroll down even more to the **Bake** menu, inside the **Render Properties** menu we covered earlier, in which we're presented with several settings:

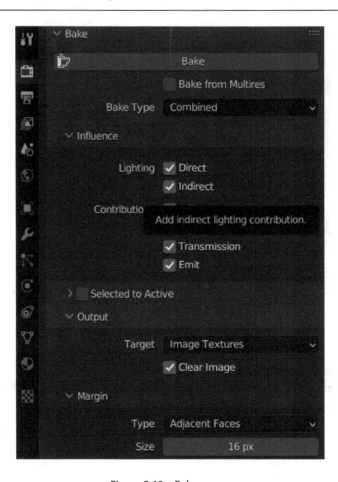

Figure 5.40 – Bake menu

From there, we'll only change three settings:

- **Bake Type**: We'll change it to **Diffuse**, so we only bake the color map.

- **Lighting**: We'll uncheck both **Direct** and **Indirect**, so the lighting in Blender doesn't affect our textures.

- **Margin**: We'll change **Size** from 16 px to 2 px, as this setting determines how much the texture will bleed from our UV islands (it's good to have a bit of bleed, to avoid any blank spots in the object's textures when displayed). Since we made all the islands close together for maximum quality, a 16-pixel bleed in our texture maps would affect multiple islands and make our object look bad. This setting also depends on the resolution of the texture, so a bigger resolution means a bigger bleed.

With all the settings changed, you can select all the objects in **Object Mode** and, with all the image textures as the active ones in all of the materials applied on all of the objects selected, click **Bake**. Now, wait a few seconds; you'll see a progress bar at the bottom of your Blender window. This is how our diffuse texture turned out:

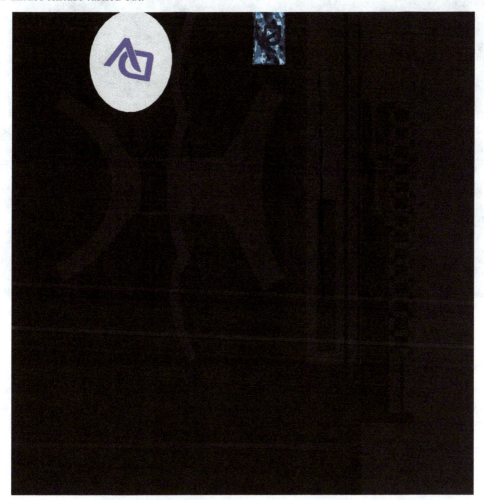

Figure 5.41 – Result of the baked diffuse map

Great, now you can save the image by selecting the **UV Editor**, selecting your baked texture, and, while hovering your cursor over it, pressing *Alt + S* to export it. If you don't want to export it just yet, you can tell Blender to pack the images internally, so the textures become embedded into the project file.

You can do that by going to the **File** menu at the top left, then going to the **External Data** menu, and finally selecting the **Pack Resources** option:

Figure 5.42 – Packing the baked textures internally

Great, now you can leave Blender whenever you want and your textures will be stored.

For the other maps, you'll do the same thing: create an image, set the resolution, add it to the materials for all objects you want to bake the textures from, change the type of map to **Roughness**, **Normal**, or whatever other map Blender can bake, select all objects in **Object Mode**, and then hit **Bake**.

Here are all the other maps we baked:

Figure 5.43 – All the other maps baked for the headphones model

Now, to test whether it really works, you can remove all the materials from the original headphones (it's recommended to make a copy of everything and save the project first) and substitute it with the new texture maps we baked only. This is our final result:

Figure 5.44 – Final result of the headphone model, with the
baked textures applied, rendered with raytracing

This looks very nice, and the high-poly model now works in most 3D programs.

Now, what if we want a low-poly version of this model, but still want to keep the detail? Well, then we'll bake all of the details from this into our low-poly version (which was made at the end of *Chapter 2*) using a very similar technique.

Baking from high poly to low poly

This method is usually used for game models and to decrease render times in the case of pre-rendered animations. It consists of baking all the textures originally applied to the high-poly model (which doesn't require fully working UVs if baking from a high-poly mesh to a low-poly mesh is the initial purpose) to the low-poly model (which *does* require fully working UVs).

The process itself is very similar to the previous, but we only need *one* material setup for the whole model, since we want every part of it to be baked into one texture set.

To start, we'll join the separate parts that compose both of our headphones (by selecting them and pressing *Ctrl + J*) into two models: one for the high poly and one for the low poly (we can separate them later). We want to make sure that they are in the *exact same place*, overlapping each other. It should look like this:

Figure 5.45 – Low-poly and high-poly model versions overlapping

It's possible to see clear signs of overlapping in this case, on the earcups, but it's not necessarily present in all cases; both of the models just need to be in the same location and rotation.

Now, go to the **Render** tab, then scroll down to the **Bake** submenu again:

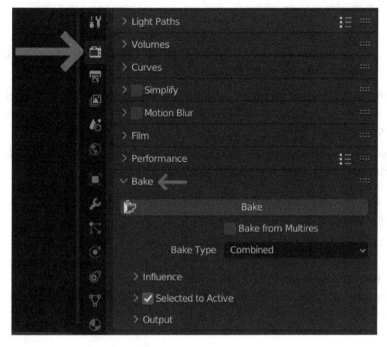

Figure 5.46 – Bake submenu

Inside there, you'll want to check the **Selected to Active** checkbox, which makes sure we're baking textures from one object to another. Then, inside that menu, we're presented with three settings:

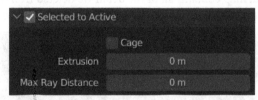

Figure 5.47 – Selected to Active menu

While most of Blender's settings try to be as intuitive as possible, these ones may not be so easy to understand by just looking at them, so here's what each one means:

- **Cage**: This is a checkbox that makes it so we can use a third object for baking the textures from one object to another (for example, use a copy of the object so that you don't have to move the target object to the same place/change the scale of the object we're baking from)

- **Extrusion**: How much the original mesh gets "inflated" before baking (this doesn't happen physically in the mesh; it's just a calculation that is necessary for the texture's projection from one object to another)

- **Max Ray Distance**: The maximum distance that "rays" are projected from one object's surface to the other object's surface for baking. 0 means unlimited

In our case, a value of 0.42 m for **Extrusion** and 0.05 m for **Max Ray Distance** worked well, giving satisfying results.

Now, like with baking from high poly to high poly, we need to have our image texture created and selected in the desired material, select what map we'll bake (**Diffuse**, **Roughness**, **Metalness**, **Bump**, **Normal**, **Emission**, etc.), then select the high-poly model first and then the low-poly model (as the second one is the active element). Finally, hit **Bake** and wait for the results. If your results don't come out looking good, though, try tweaking the settings and test again, making sure to save the desired image when you get a satisfying result.

After the first map is baked, you can connect it to the correct input if you haven't already and move on to the next map, adding another image texture, setting up a new image, selecting it, baking, and repeating until all of the desired maps are baked.

You'll notice that the normal map contains all of the detail we deleted from the original, high-poly model, and that was the objective. See the following comparison:

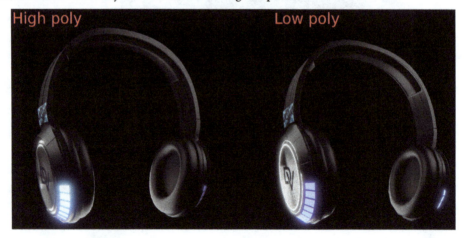

Figure 5.48 – High poly versus low poly final results comparison

With that, we have successfully baked the high-poly detail into the low-poly model, as you can see. The final result should work better as a game model, for example, since it has about four times less geometry to render. Now, we can call this model fully finished.

Summary

In this chapter, we covered two more texturing techniques and how to apply them in different situations. We also covered how to bake the textures in order for the model to work outside of Blender.

But what if you don't want to make inorganic assets? What if you're more interested in making natural elements, such as trees, cliffs, or rocks? Or, what if you want to model characters? Regular poly modeling would take much, much longer. Well, the next chapters will be perfect for you, as they cover in detail the advantages of the sculpting features in Blender.

Part 2:
Organic Asset Modeling

In this part, we'll go over the process of creating inorganic assets using the sculpting tools in Blender, going over most of the sculpting brushes available by default, explaining how they behave, and suggesting use cases for each. In this part, we'll sculpt two distinct organic objects, and we'll also study human anatomy in order to sculpt and optimize a full-body human.

This part has the following chapters:

- *Chapter 6, Introduction to Blender's Sculpting Tools*
- *Chapter 7, Making the Base Mesh for a Humanoid Character*
- *Chapter 8, Refining the Base Meshes*
- *Chapter 9, Optimizing the Base Meshes*
- *Chapter 10, Rigging the Base Meshes*
- *Chapter 11, Further Development as a 3D Artist*

6

Introduction to Blender's Sculpting Tools

If you want to do anything organic, sculpting is your best bet, as sculpting allows us to tweak more than a small set of vertices/faces at once in several different ways.

Blender offers a powerful sculpting system, with several different brushes that affect the surface in different ways, making it possible to achieve a huge amount of variation and different effects in our mesh.

In this chapter, we'll cover the following:

- The default Blender sculpting brushes and their settings
- The advantages that using a drawing tablet brings when sculpting
- Suggested situations for when to use each brush and when to combine brushes

By the end of this chapter, you should be familiar with most of the sculpting tools available by default in Blender.

Blender's default sculpting tools

Before we start any sculpting, we need to know what tools we have to do the job, right? So, we'll start by going into each of the main UI elements, the default brushes available, and their main settings.

UI elements

To access the **Sculpting** workspace, we can either select it from the menu at the top-left of the screen (as we covered while texture painting) or select the **Sculpting** tab from the several options in the menu bar above the viewport (third from left to right).

Here's the full window once you enter **Sculpt Mode**:

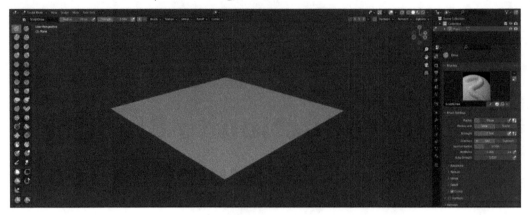

Figure 6.1 – Full sculpting workspace window

We'll cover all the interface elements of this mode (except the brushes themselves, which will have a dedicated section right after this one) from left to right, starting with the top bar of settings.

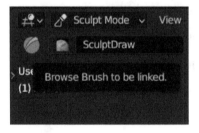

Figure 6.2 – Active brush

At first, we have the actual brush we're using, with its name and then its thumbnail to the right of the name. Besides showing the name of the active brush, it also lets us select from the custom brushes we assigned to it, which won't be covered, though, as we'll focus on the default sculpting tools.

To its right, we have the **Radius** and **Strength** sliders, which control the radius of the stroke applied and the intensity of the effect that the said stroke will have on the surface:

Figure 6.3 – Radius and Strength sliders

Although you can use the sliders to control those parameters, it's faster to use the shortcuts *F* and drag your cursor left or right to control **Radius**, and *Shift + F* and drag to control **Strength.**

To the right of each slider, there's an option that activates or deactivates pressure sensitivity, which only makes a difference when using a graphics tablet (the advantages of doing so will be covered along with the brushes).

Next, we have the **Add** and **Remove** buttons, which control whether the brush will add volume to the surface or carve into it:

Figure 6.4 – Add and Remove buttons

Then we have the **Brush** menu, which contains several settings and sliders to control how the active brush behaves:

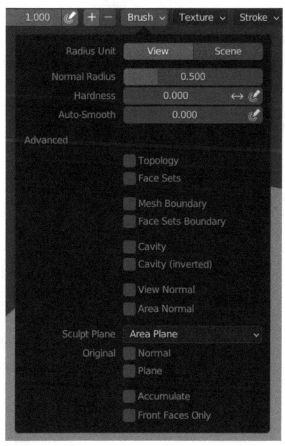

Figure 6.5 – Brush menu

Out of all the settings available, **Hardness**, **Topology**, **Cavity**, **Accumulate**, and **Front Faces Only** are the ones that can make the most difference in the stroke:

- **Hardness**: Controls how close to the edge of the brush the falloff starts.

- **Topology**: Controls whether or not the brush affects unconnected vertices within the same object.

- **Cavity**: Makes it so that the stroke doesn't affect vertices on peaks or valleys, based on surface curvature.

- **Accumulate**: If activated, when one brush stroke overlaps with itself, it applies that stroke again with the same intensity, adding more volume. This can create harsher surfaces.

- **Front Faces Only**: Since the brush radius is spherical, this controls whether the brush stroke affects the vertices located behind the surface on which the brush is being applied. Can be useful when sculpting on thin surfaces.

If you forget or want to know what any setting does, you can hover your cursor over it and a little explanation will appear.

Next, we have the **Texture** menu, which can be used to add custom textures:

Figure 6.6 – Texture menu

As has been said already, we won't cover this feature in depth, but it works similarly to when we added custom brushes while texture painting, using black and white textures.

To its right, we have the **Stroke** menu, which controls how the stroke will be laid out:

Figure 6.7 – Stroke menu

This menu works exactly how it works in texture painting, presenting options to control things such as the spacing in between the application of the brush to the surface, jittering of the brush's rotation during the strokes, and whether the stroke gets stabilized or not, in order to create a smoother, possibly less organic stroke.

Then, we have the **Falloff** menu, which controls how the brush's effect fades out when it gets closer to the edges of the brush:

Figure 6.8 – Falloff menu

Inside that menu, we have a submenu with many presets, as well as the option to make our own custom falloff:

Figure 6.9 – Falloff presets

Next, we have the **Cursor** settings, which allow us to modify the appearance of our cursor while sculpting. These are only appearance settings and won't affect the sculpting process in any way.

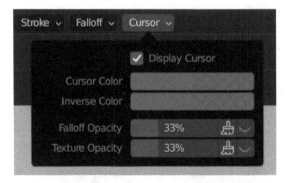

Figure 6.10 – Cursor menu

Now, on the top-right bar, we have the **Symmetry**, **Dyntopo**, and **Remesh** menus:

Figure 6.11 – Top-right bar of settings

- **Symmetry**: Makes the strokes symmetrical in one or more axis as they are made. You can also symmetrize the entire object at once if you access the **Symmetry** menu, as well as other settings such as radial symmetry.

- **Dyntopo**: Adds geometry as the strokes are made, depending on a set level of detail, which can be defined in its menu.

- **Remesh**: Substitutes the existing geometry with quads only, based on a set level of detail, which can be set by changing the **Voxel Size** slider inside the menu or by using the shortcut *Shift + R* and dragging your cursor, and *Ctrl + R* to apply the remesh.

All of the options and settings presented so far can be accessed through the **Active Tool** tab as well:

Figure 6.12 – Active Tool menu

It's worth noting that some brushes may have additional or fewer settings, depending on how they work, and that the settings are brush-specific – that is, if we set something for one brush, it won't be applied to any other.

Alright, now we know what tools we have to control our brushes, but what about the brushes themselves? Let's take a look at the most important brushes.

Blender's sculpting brushes

Now, with the knowledge of the general settings, we can go ahead and go over the most used brushes when it comes to sculpting (we'll skip some that are not so common, but you can try those for yourself to see how they behave). We'll go from top to bottom, and after each explanation, there will be an image demonstrating each brush.

Note that, for sculpting, we need a very large number of faces in order to display a proper amount of detail. For instance, the plane we'll use for the demonstrations has around 65k faces.

It's highly recommended that you play around with the different brushes, settings, and shortcuts, as this helps to better understand how each thing works and behaves.

Figure 6.13 – Brushes menu

At the top, we have the default **Draw** brush, which pushes the surface outward or inward (depending on whether you're holding *Ctrl* while using it, or have selected the **Add** or **Remove** parameters), based on the average of the normals of the faces under the cursor; it can also be accessed using the shortcut *X* while in **Sculpt Mode**.

We can apply the brush stroke by left-clicking and dragging with a brush selected.

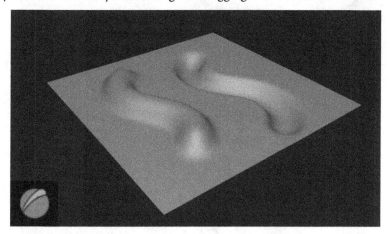

Figure 6.14 – Draw brush

Next, we have the **Draw Sharp** brush, which is, essentially, the Draw brush with a sharp falloff (its default mode is pushing inward, though):

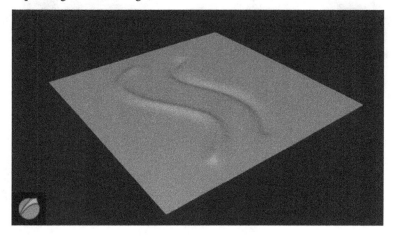

Figure 6.15 – Draw Sharp brush

Thirdly, we have the **Clay** brush, which works similarly to the **Draw** brush as well, but leaves a flatter, smoother top, a more subtle bump, and slightly sharper edges:

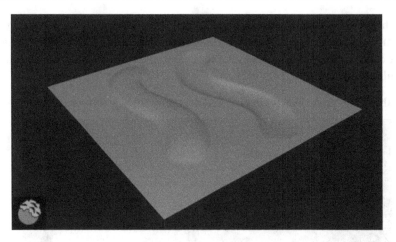

Figure 6.16 – Clay brush

Next, there's the **Clay Strips** brush. It has a square shape (you can adjust the roundness of the brush in the **Brush** menu, though), a sharp falloff, and harsher edges, and it accumulates more intensely:

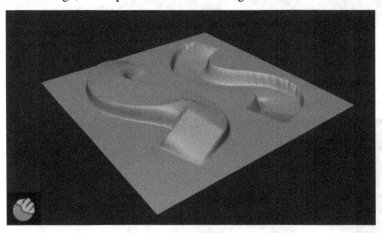

Figure 6.17 – Clay Strips brush

Then, there's the **Clay Thumb** brush, which simulates a thumb being dragged across a clay surface, so it carves a little into the mesh while building up the volume at the end of the stroke, and as it progresses, the stroke gets deeper. This brush doesn't have **Add** and **Remove** options.

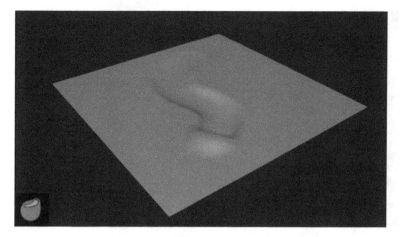

Figure 6.18 – Clay Thumb brush

Next, we have the **Layer** brush, which can be accessed using the shortcut *L*. It works similarly to the **Draw** brush, but with a slightly sharper falloff. It adds a layer with a flat top, as the name suggests, at a set height. If a single stroke overlaps itself while being made, no additional volume will be added or removed. You can add a layer on top of another, though you'll have to start another stroke.

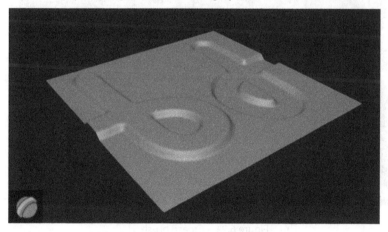

Figure 6.19 – Layer brush

Moving down, we have the **Inflate** and **Blob** brushes, respectively. They work in very similar ways, but while the **Inflate** brush moves all the vertices it gets applied on in the average normal direction, the **Blob** brush pushes each vertex in its respective normal direction, effectively making a more spherical shape.

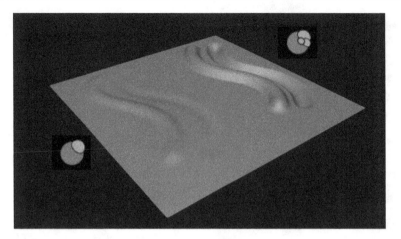

Figure 6.20 – Inflate (left) and Blob (right) brushes

Then, we've got the **Crease** brush, which works similarly to the **Draw Sharp** brush, but has a smoother falloff. It also pushes the vertices slightly toward the center of the brush to create a sharper crease (works best with smaller brush sizes).

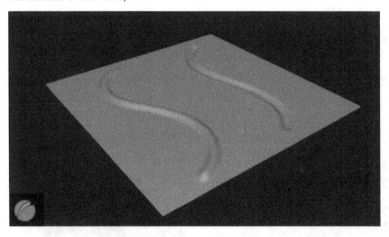

Figure 6.21 – Crease brush

Next, we have the **Smooth** brush. It averages the distance between the vertices in order to create a smother surface. This brush is so commonly used that it can be accessed while using any other brush by holding *Shift* while making the stroke (for the demonstration, a symmetric rocky terrain-like surface will be sculpted using the **Clay Strips** brush, and half of it will be smoothed out using the **Smooth** brush). Holding *Ctrl* to invert the action accentuates the irregularities.

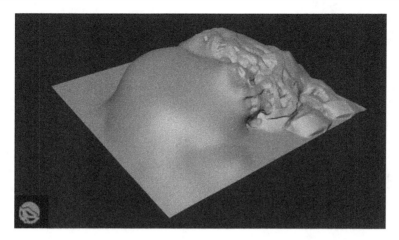

Figure 6.22 – Smooth brush applied to half of a symmetric sculpt

Moving on, we see the **Flatten** brush, which calculates the average height of the vertices under its influence area and moves the ones above and below it to that height. As the name suggests, it flattens the surface it's applied on (again, we'll apply the brush on half of a symmetric sculpt for the demonstration). It can also be activated using the shortcut *Shift + T*.

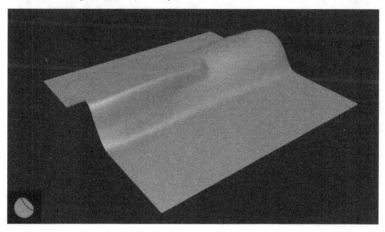

Figure 6.23 – Flatten brush applied to half of a symmetric sculpt

Next is the **Fill** brush. It works similarly to the **Flatten** brush except it only moves the vertices below it to the average height calculated under the brush's area of influence (the area of calculation, though, can be adjusted without affecting the radius of the actual brush, in the **Active Tool** menu or in the **Brush** settings).

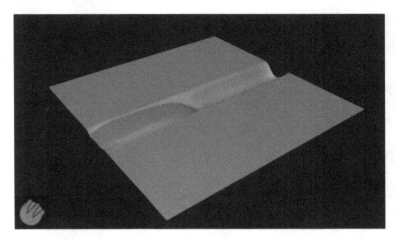

Figure 6.24 – Fill brush applied to half of a symmetric sculpt

Then we have the **Scrape** brush. It is the opposite of the **Fill** brush, meaning it moves down the vertices above the average calculated height.

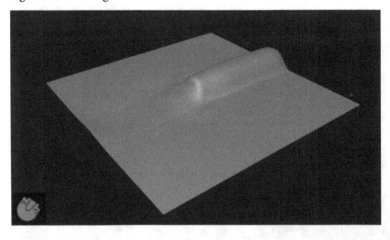

Figure 6.25 – Scrape brush applied to half of a symmetrical sculpt

Below it, we have the **Multi-Plane Scrape** brush, which works by using two planes at a set angle (which can be adjusted dynamically or manually, depending on the settings we use) and applying the **Scrape** brush on the area of influence of those planes (the demonstration was made using a fixed, manually set angle of 60°, and you can see the two planes used, which is the brush, at the end of the stroke).

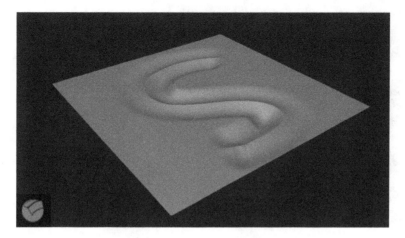

Figure 6.26 – Multi-Plane Scrape brush

Then, we have the **Pinch** brush, which moves the vertices under the brush to its center, tightening the crease or peak to make it sharper. It also pinches the geometry on flat surfaces, but it doesn't make any difference visually. It can also be activated using the shortcut P.

Figure 6.27 – Before (left) and after (right) of the Pinch brush being applied to a peak and a crease

Next is the **Grab** brush. It selects the vertices under the brush cursor and drags them, following the cursor's position. It's similar to the proportional editing tool in edit mode. It can be activated using the shortcut G.

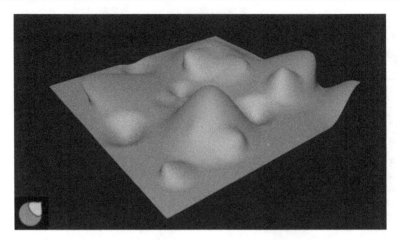

Figure 6.28 – Grab brush used to drag up and down multiple spots

Next, we have the **Elastic Deform** brush. It works very similarly to the **Grab** brush, by dragging the vertices with the cursor, except this brush simulates an elastic surface while doing so, therefore creating a more natural and smoother deformation.

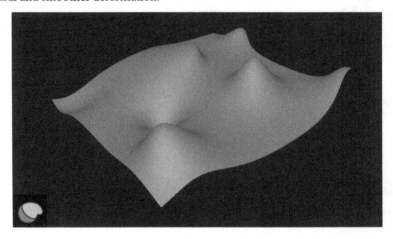

Figure 6.29 – Elastic Deform brush

Moving down, there's the **Snake Hook** brush, which works like the previous brushes but it moves the geometry along the stroke, leaving a "path" behind.

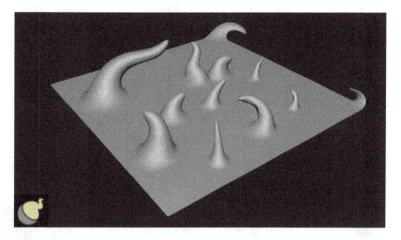

Figure 6.30 – Snake Hook brush

As you can see, the tip of each stroke got rotated in the direction of the stroke, and this can be adjusted in the **Brush** settings, as well as the loss of volume as the stroke gets longer.

Keep in mind that this doesn't add more geometry; it pulls the existing geometry, so expect a loss of resolution as the stroke gets longer – that is, if you don't enable **Dyntopo** before applying the brush, which tends to work and look better for longer strokes of this brush.

Now, we'll start to skip a few less frequently used brushes, as we're mostly focusing on the more useful ones.

The next big brush is the **Pose** brush. It allows us to easily deform our sculpt as if there were bones and joints there. It's useful for posing the limbs of characters you don't want to animate or to change poses frequently.

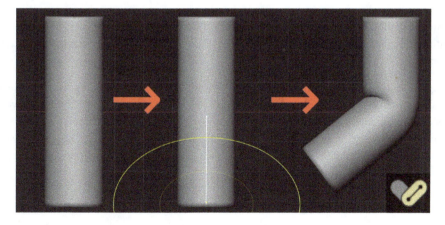

Figure 6.31 – Pose brush

By default, this brush works with an estimation made by Blender based on the current geometry, so it may get the deformation wrong. We can tweak the settings until we get a satisfying result if that happens.

Now, we have the **Slide Relax** brush, which slides geometry without affecting the volume of the sculpt itself, which is useful to move more geometry to places where more detail is needed without actually increasing the number of faces we have on the mesh itself. It's useful on smaller creases and peaks (for the demonstration, we'll turn on the wireframe overlay, so that we can better see the effect of the brush).

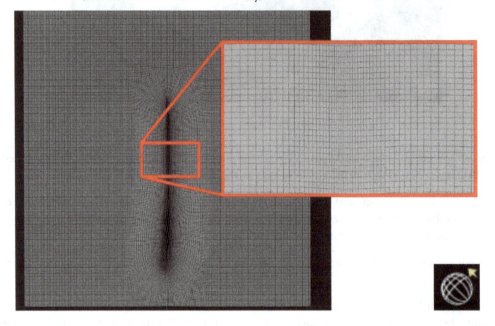

Figure 6.32 – Slide Relax brush

Notice how the geometry got dragged to the center of the stroke, allowing that area to have more details without increasing the number of faces in the object. Holding *Ctrl* while applying the stroke does the opposite.

The next big brush is the **Cloth** brush, which applies a localized cloth simulation to the area influenced by the brush (you can also apply a **Global** simulation in the settings, controlled by the brush; we'll use **Global** for the demonstration). This brush is especially useful to make cloth objects without having to set up an entire simulation or sculpt the wrinkles/creases by hand. You can also adjust several parameters of the cloth simulation, such as the cloth mass, damping, simulation area, and force falloff.

This brush is also really fun to play with.

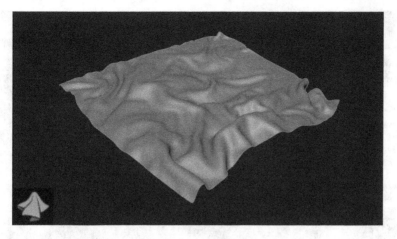

Figure 6.33 – Cloth brush

Note that, currently, we don't have any way of controlling the size of the wrinkles, but we can get around this limitation by adding or removing geometry: less geometry means bigger wrinkles, while more geometry means smaller ones.

And last but not least, we have the **Mask** brush. It allows us to "paint" an area that we don't want any brush to affect, allowing us to have more control over what we actually affect while making the strokes.

Figure 6.34 – Mask brush

The gray area delimits where the mask was painted. If we then apply any brush, let's say the **Cloth** brush, that area will remain intact (the light gray areas will be affected by the brushes, however, they will have a significantly lesser effect than something unmasked).

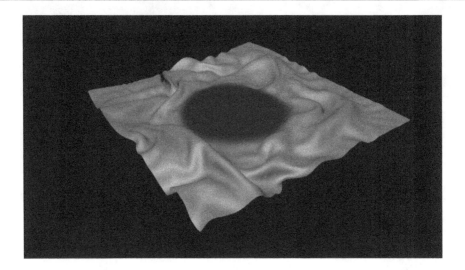

Figure 6.35 – Masked area after applying the Cloth brush

If we now remove the mask with the shortcut *Alt + M*, we'll notice that the area was not affected by the cloth simulation:

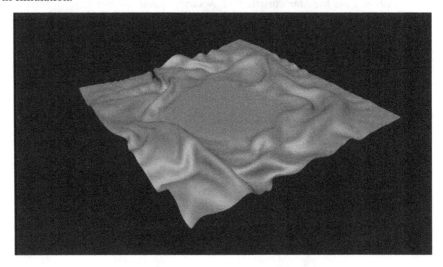

Figure 6.36 – Mask removed

We can also invert the mask with *Ctrl + I* while in **Sculpt Mode**.

Additional masking modes are also available if we scroll down the brushes column – **Box Mask**, **Lasso Mask**, and **Line Mask** – which you can access by clicking and holding the **Box Mask** icon:

Figure 6.37 – Additional masking methods

Keep in mind that we didn't cover every single brush available by default in Blender, and there are many more, less common brushes that you can test out yourself. For the rest of this book, though, we'll use only the ones listed here. You can check out the following link to understand the other brushes: https://docs.blender.org/manual/en/2.80/sculpt_paint/sculpting/tools.html.

You should also note that when sculpting, things get very organic looking very quickly, so you should avoid sculpting on things that are not supposed to look that way – at least when you're starting out.

Now, if you've already played around with drawing programs on your computer, you may have noticed that it's pretty difficult to draw using a mouse. That difficulty also translates to sculpting, which can make it more difficult to get good results. Just like with drawing on a computer, though, we *can* get good results when sculpting with a mouse; but again, just like when drawing, using a graphics tablet generally makes the task easier.

Advantages of using a graphics tablet to sculpt

A drawing tablet offers us a variety of advantages while sculpting and/or drawing on a computer.

A regular mouse was made for precision and sharper movements, but generally, sculpting doesn't require that amount of precision, since most of the time, we're trying to make things look natural and organic. That's where a drawing tablet comes in handy.

It offers a great amount of control for our strokes by allowing us to use a paper-like surface to "draw" on and most importantly, emulates actual drawing motions, as they come with a pen as well. Along with the more natural motion, we also get control over the strength of the brushes by adding physical pressure to the pen, which adds a whole other level of detail to our sculptures that would be much harder to achieve using a mouse.

Fortunately, most of the 3D modeling/sculpting software today offers support for drawing tablets, and Blender is no exception.

To enable the said feature, we need to find the following icon and activate it:

Figure 6.38 – Strength Pressure button

Upon activating that setting, the icon will turn blue, and whatever parameter it controls will be influenced by the pressure applied against the tablet by the pen.

Pressure can control parameters such as brush radius, strength, and opacity when using the **Mask** brush or texture painting.

When you connect a tablet to use while sculpting, though, you may notice that we gain control over the strokes we make, but we lose control over the viewing angle, which can't be easily changed with a tablet as we no longer have the scroll wheel, and it would be very time-consuming, not to say annoying, to switch from tablet to mouse every time we want to move our camera around. To solve this, we'll have to tell Blender to emulate a three-button mouse.

We'll start by going over to the **Edit** menu at the top left, and selecting the **Preferences** tab:

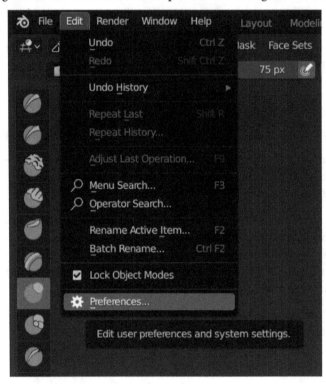

Figure 6.39 – Accessing the Preferences tab

Now, the **Blender Preferences** window should pop up. Inside there, we'll head to the **Input** tab on the tab menu to the left of the window, where we'll find different options and settings for how Blender interprets the keyboard and mouse inputs. The one we're looking for is the **Emulate 3 Button Mouse** option. Check its checkbox.

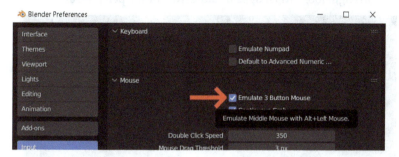

Figure 6.40 – Emulate 3 Button Mouse setting

Now, whenever we want to move our view around, pan it, or zoom in or out, we can hold *Alt*, *Alt + Shift*, or *Alt + Ctrl*, respectively, and then drag the pen along the tablet.

This trick works for laptop touchpads as well, and also for mice that don't have a scroll wheel for some unknown reason.

From now on, all our demonstrations when it comes to sculpting will be made using a drawing tablet, unless otherwise stated. The tablet we'll be using is the Huion Kamvas 13, although simpler tablets can get you similar or identical results. It's all a matter of getting comfortable with the tools available.

Let's see how the pressure control looks in practice, starting with **Radius**, then **Strength**, and lastly, both combined.

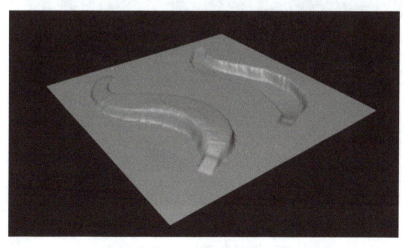

Figure 6.41 – Clay Strips stroke applied with the brush radius controlled by pen pressure

Notice how the start and end of each stroke are thinner than the middle, while the strength remained the same. This can make tasks such as sculpting pointy things (for example, sticks and horns) a lot easier. You can still set the maximum radius using the slider.

Now let's see how **Strength** looks when dynamically driven by the pen pressure.

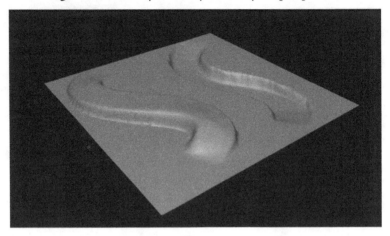

Figure 6.42 – Clay Strips stroke applied with the brush strength driven by the pen pressure

As you can see, both strokes have smoother starts and finishes, while the middle section carried full strength (set by the value in the slider) when the pen was pressed down with more force. This can make blending different surfaces much easier, for example.

When both aspects are controlled by the pen pressure, the stroke looks like this:

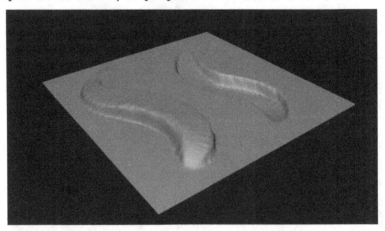

Figure 6.43 – Clay Strips brush stroke applied with both Radius
and Strength being driven by the pen pressure

Alright, now that we know about most of the sculpting tools we have to hand, we'll go over some of the use cases for a few of these brushes, as the demonstrations alone up to this point may not make them apparent.

Suggested uses for the brushes

None of the brushes have a "correct" use case, so these are just suggestions on how the shape and functionality of some brushes can be used to the artist's advantage, and additionally, to understand how the brushes mentioned here can be used in realistic scenarios.

Let's start with the rounder, smoother brushes such as the **Draw**, **Clay**, **Inflate**, and **Blob** brushes.

These build up the volume in a spherical manner, so they result in a better appearance when used in situations where, say, liquid runs and/or accumulates, and to show round organic objects and fruits, fat accumulation and veins on characters, and so on, and so forth.

Let's have a look at candles, for example:

Figure 6.44 – Candle references (images from https://www.pexels.com/)

As you can see, the molten wax accumulates in the borders near the flames and at the bottom, as well as a few running drips down the side. Notice how they accumulate in a round and soft-looking way.

We can replicate that using these rounder brushes, starting with a cylinder, remeshed using *Shift + R* and *Ctrl + R*, to add more geometry (we set the voxel size to a bigger number initially, to block out the bigger shapes such as the "hole" where the fire is melting first and the bigger deformities along the body of the candle):

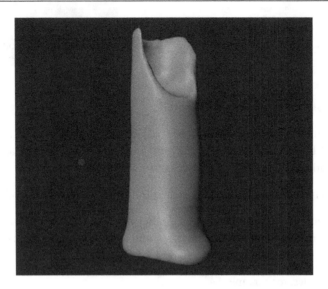

Figure 6.45 – Defining the candle's general silhouette

See how most of the candle has a round, soft-looking, organic surface. This look is easier to get with brushes that deform the mesh in a rounder way.

Good, now that we're satisfied with the general silhouette (which was made using the **Draw** and **Smooth** brushes only to this point), we can remesh it again with a smaller voxel size to add more detail. This time, we'll puff up the borders and add a few bigger drops and wax paths where they tend to form more frequently and accumulate:

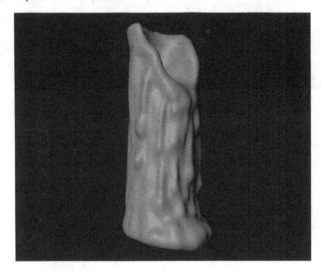

Figure 6.46 – Bigger molten wax drops added

Notice how some of the drops are more subtle than others, especially when they get further away from the main section where the wax was melted more, and that this part has more drops.

The drops and the puffed-up border were made using the **Draw**, **Inflate**, and **Blob** brushes, and by using the **Front Faces Only** brush setting, as the molten borders are very thin and the opposite side could be affected by a brush stroke with spherical falloff.

For some people, this would show enough detail for a candle, but it's possible to go further by adding geometry and making smaller drops:

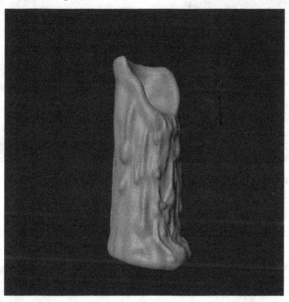

Figure 6.47 – Smaller wax drops added

A higher level of detail is only necessary if the object is closer to the camera, or if the model itself is being made as a study. The candle wick that is lit could be made using simple modeling.

Remember that the more detail you need or want, the more geometry will need to be added. This candle, for example, has about 140k faces at its last stage, with more detail. Right now, though, optimizing these objects is not our focus, but rather understanding how to use some of the brushes in order to get the best-looking results out of them.

Now, let's see how the sharper-ended brushes can be used, such as the **Draw Sharp**, **Clay Strips**, **Crease**, and **Pinch** brushes.

They can be used in a variety of ways, but it's common to see them being used for sculpting surfaces with a rougher look such as rocks, tree trunks, sticks, horns and antlers, skin wrinkles, and so on – things that need a harsher/sharper appearance in general.

If we look at bigger rocks (on mountains, for example), we notice that they are usually much less smooth than those found in a river or lake, and that rough look can be easily replicated using these brushes.

Figure 6.48 – Rough rocks (image from https://www.pexels.com/)

We can start by either using a very low poly mesh to define the general silhouette, usually using the **Grab** brush, which is great for that, or we can use a few cubes to do it, which generally works better for things like this since the sharp edges create a harsher look. We'll stick to the cubes:

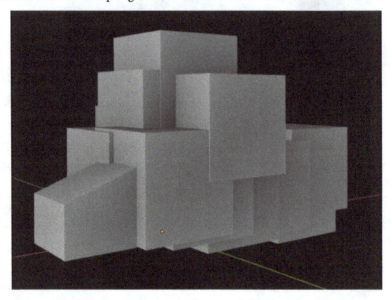

Figure 6.49 – General rock silhouette defined using cubes

Notice how most of them remained as cubes, overlapping and with no editing whatsoever. This is because we'll do all the rest of the shaping using sculpting.

We can now join all the cubes using *Ctrl + J* and remesh the resulting object with a larger voxel size to get rid of the overlapping and to join their surfaces into one so that we can sculpt on all of them at once to add the bigger, more defining details.

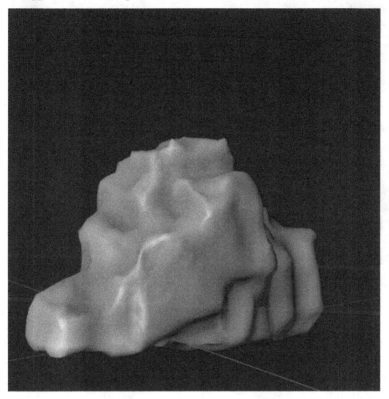

Figure 6.50 – Larger details sculpted

This iteration was made using the **Clay Strips** brush only, as there is still insufficient geometry to display proper detail for smaller brushes (3.5k faces).

Now, with a general shape we're satisfied with, we can move on to smaller details, remeshing the rock once again with a smaller voxel size (you can also use **Dyntopo** and progressively increase the amount of detail):

Figure 6.51 – Smaller details added

Now, with more geometry (45k faces), we were able to add finer details using mainly the **Clay Strips** brush, along with **Draw Sharp** on the peaks set to a larger radius to define the edges and creases a bit more, as well as the **Scrape** and **Smooth** brushes to flatten some of the areas that are not supposed to look as rough or pointy.

We can go further and add even finer details:

Figure 6.52 – Finer details for the rock

Notice how some of the edges have been accentuated, using the **Draw Sharp** brush in the **Add** mode on the peaks, and in the **Remove** mode on the cracks at a lower strength, as well as the **Pinch** brush to make the cracks tighter and sharper-looking (which can also be done on the peaks). This makes the peaks look more pronounced.

From now, if you want to add even more detail, a custom, rougher brush might help a lot, as all of the little cracks, indents, and other details tend to be extremely hard to add manually, one by one:

Figure 6.53 – Further details added using custom texture for the brush

This step is optional most of the time, though, since a lot of the details can be added afterwards with texturing and baking. Again, the final face count is high, around 176k faces. To get smaller details in, you'd most likely need to reach millions of faces, which can be done, but for that, it's recommended to use a **Multiresolution** modifier instead of remeshing, so that you can easily go back to less detail if needed.

We recommend practicing by sculpting a few more organic objects to get a better grasp and fluency at using the brushes, so here are a few more challenges to do for yourself: a fruit, such as an apple or pear, which has a more smooth and rounded shape to it. Next, you can try sculpting mushrooms, since they require a slightly more complex block out and have more intricate shapes, sometimes having a mix of smooth and harsh edges. Finally, you can try sculpting simple shells, such as seashore and/or snail shells, which will make you better at making relatively smooth surfaces with creases, especially with snail shells. Don't forget to gather references, though, as it is the key to instantly making your work better.

Perfect! Once you feel comfortable using most of the sculpting tools and know how the main brushes behave in realistic situations, you can move on to a more complex subject: characters. They present unique challenges when sculpting and those can be tricky to get around, especially for a beginner or someone without any sort of understanding of human proportions and anatomy. This is what we'll cover in the next chapter.

Summary

In this chapter, we covered the main sculpting brushes and settings available by default in Blender, with examples of their uses in practical examples, as well as how to progressively add details to a sculpt. You should now be familiar with most of the sculpting tools in Blender, although practicing further is highly recommended.

Now, we'll see how to approach the creation of a character, as it differs from organic objects in quite a few aspects.

7
Making the Base Mesh for a Humanoid Character

Unlike making organic objects, sculpting characters presents us with a series of new things to worry about, especially if the end goal is to resemble a human and look believable. This chapter is essential if you plan on making characters in the future, as it will cover the most important features of the human body and how to block out a very simple depiction in order to have a solid base to which to add detail later on.

For that reason, this chapter will cover the following:

- General human body proportions and anatomy, which make a significant difference while sculpting

- How to block out a base mesh for a humanoid character, using realistic proportions and shapes

By the end of this chapter, you'll have learned about basic anatomical features and proportions and will be able to block out humanoid male and female base meshes using realistic proportions.

General body proportions for humanoid characters

When dealing with humanoid characters, the initial stages are crucial to achieving a believable end result, and understanding realistic body and face proportions are a big part of that, even if in the future we'd like to make stylized characters with exaggerated proportions. It's important to learn the rules before we break them.

We'll start by taking a look at an anatomical model of a full human body for general proportions, and then proceed to break down the main parts, such as chest, arms, legs, hands, and feet. We'll look at the main structures that shape each part (mainly muscles).

For learning purposes, we'll work with general body proportions:

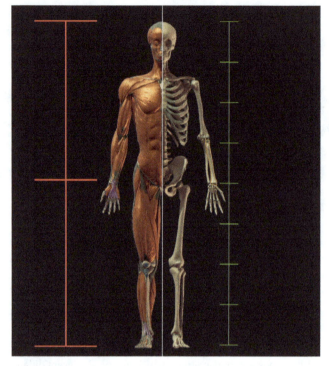

Figure 7.1 – Full body anatomical human model (model provided by the Z-Anatomy project at https://www.z-anatomy.com/)

We'll start by analyzing the height of the model and how it can be determined for a character.

Using head/body proportion (the green markers in *Figure 7.1*), the average height of a human is around 8 heads tall. And although this can vary from person to person, usually, this variation is only around half of a head, leaving us with a height that can vary from 7.5 to 8 heads tall for a fully grown adult. This proportion changes as we grow, though. Here are a few other age groups and their respective ideal head/body proportions:

- Infant (~1 year): Around 4 heads tall

- Child (~3 years): Around 5 heads tall

- Child (~5 years): Around 6 heads tall

- Child (~10 years): Around 7 heads tall

- Teenager (~15 years): Around 7.5 heads tall

Another thing worth noting for this first analysis is the leg/body proportion (red markers), which is half, meaning that our legs usually make up around 50-60% of our height. This also indicates at which height the groin area is usually located: at around half of the height. Now, we can start going deeper into the proportions of each of the main limbs and the face, which will be covered at the end.

Arms and hands

We'll start by analyzing the proportions of the arm, its main muscle groups, and how they interact, as well as how they affect the overall shape of an arm.

For that, we'll have a look at the left arm:

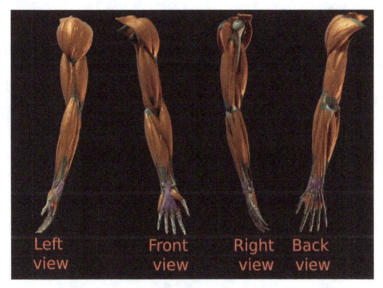

Figure 7.2 – Left arm (model provided by the Z-Anatomy project at https://www.z-anatomy.com/)

As for proportions, notice that the length from the shoulder to the elbow is nearly identical to that of the elbow to the wrist, putting them at around a 1:1 ratio, relative to each other. This can be observed very easily by bending your arm as much as you can and seeing how close your wrist gets to your shoulder.

To determine the size of the arm relative to the body, though, remember that with a relaxed arm, the elbow nearly lines up with the belly button (or three heads down the character's height), sometimes sitting right above it, and the hands can almost reach the middle of the thighs as well while standing.

As a rule of thumb for arm/body proportions, remember that our hands can comfortably reach down to our crotch. A lot of beginners tend to make the arms too small when dealing with realistic proportions.

Now, let's talk about the actual shape of an arm, mainly defined by its muscles. The arm appears to be composed of three main muscle groups:

- **Shoulder** (tinted purple in *Figure 7.3*): The shoulder is mainly shaped by the **deltoid** muscle, which gives it that rounded contour. Anatomically, though, this muscle appears to be composed of three different sets of fibers, but this separation doesn't usually influence much in the shoulder's shape.

- **Arm** (tinted green in *Figure 7.3*): The middle section (which goes from the shoulder to the elbow) is shaped by the **biceps** and **triceps**, which are large muscles and extend from the shoulder muscle to the elbow. The biceps sit at the front and are responsible for flexing the arm, while the triceps are located in the back part and straighten the arm. They also have several smaller muscles that make less of a difference in the actual shape of the arm.

- **Forearm** (tinted blue in *Figure 7.3*): This group is made up of several muscles, but the ones that affect the shape more are the **brachioradialis, flexors**, and **extensors**, which are responsible for flexing the forearm, the contraction, and extension of the wrist, respectively. The brachioradialis muscle and the flexors sit at the front of the forearm while the extensors sit at the back. They extend from the elbow all the way to the wrist, getting thinner as they get closer to the hand. At the wrist, some of the flexor muscles can be seen as tendons (*Figure 7.4*).

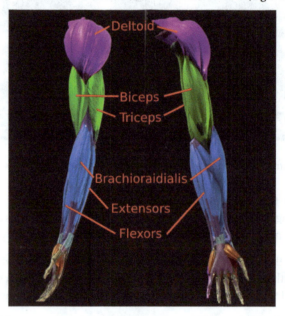

Figure 7.3 – Muscles of the arm separated by visual groups (model provided by the Z-Anatomy project at https://www.z-anatomy.com/)

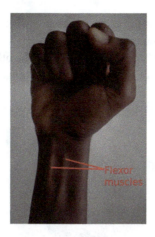

Figure 7.4 – Flexor muscles are seen as tendons at the wrist (image from https://www.pexels.com/)

Now, let's have a look at a hand and analyze its proportions and shape:

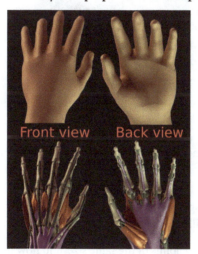

Figure 7.5 – Left hand (model provided by the Z-Anatomy project at https://www.z-anatomy.com/)

To analyze relative proportion, we'll use the palm/finger ratio, which is measured by the height of the middle finger, as it is the longest. As you can see, it is close to 1:1, with the middle finger usually being slightly smaller than the palm (*Figure 7.6*, red markings). For the other fingers, we can use an arch to determine the general curvature present when comparing the fingertips (blue marker):

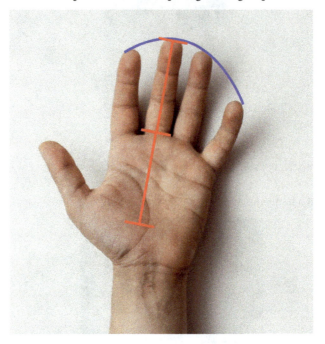

Figure 7.6 – Arch formed by the fingertips and palm/finger proportion (image from https://www.pexels.com/)

These proportions can vary from person to person, so there's generally a bit of room for testing hand proportions overall with them still looking believable.

Now note how each finger has three joints of the same size, and how the thickness of all the different fingers is nearly the same, with very little difference in each finger's thickness lengthwise as well.

The thumb can be a little trickier to deal with due to its location on the hand and its size, but it generally follows the same shape as the other fingers and its tip usually reaches the lower part of the index finger (when put against the side of the palm). When relaxed, it tends to form a 30° to 45° angle relative to the palm.

Now, let's talk about shape. There are three main muscle groups that give our hands their shape: near the thumb (tinted purple in the following figure), near the index finger (tinted green), and near the pinky (tinted blue). The rest is mostly bone, tendons, and fat tissue:

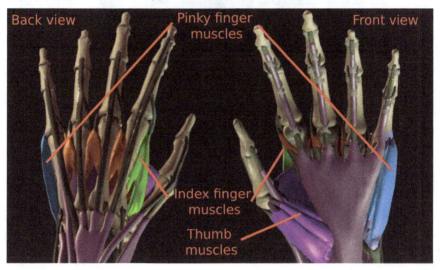

Figure 7.7 – Main hand muscles that shape the left hand (model provided by the Z-Anatomy project at https://www.z-anatomy.com/)

The thumb muscle group is the more pronounced in terms of visibility, being the most "puffed up" area of the hand, near the wrist. The pinky finger muscles are also pronounced but are located on the side of the palm, allowing for a tighter grip on the pinky finger. The index finger muscle group is located on the back of the hand and is most noticeable when flexing that finger.

Also, note that there are no actual muscles in the fingers themselves. This makes it so that the front and back parts of the finger look a bit flatter than the sides.

A good way to help grasp how the hand is shaped is to look at your own. Move it, rotate it, touch it, see how the muscles react to the movement, and observe the overall proportions.

Now, with the upper limbs covered, let's analyze the lower limbs.

Legs and feet

The legs are the largest limbs on the human body, and it is important to get the scale and relative proportions right to correctly convey the height of the character. Let's start by having a look at a left leg:

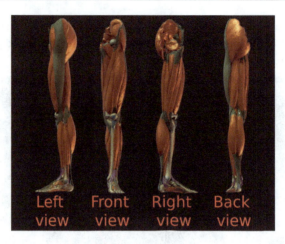

Figure 7.8 – Left leg (model provided by the Z-Anatomy project at https://www.z-anatomy.com/)

The leg is around four heads tall, and as you can see, the lower leg is a bit smaller than the upper leg, although a 1:1 proportion can also work well in some cases. Note as well that the leg doesn't go straight down but rather at a slight angle.

Looking at it from the front, the widest area of the leg usually is the upper part of the thigh, or near the waist in a few cases. In contrast, the slimmest part of the leg is at the ankles, where there aren't many big muscle groups.

Speaking of which, a leg seems to have four main groups visually, as follows:

- **Gluteus muscles** (tinted pink in the following figure): These are located at the back of the leg above the hips and are responsible for extending the whole leg backward and/or pushing the pelvis forward, as well as assisting with balance. There are two main muscles in this group – the **gluteus medius** and the **gluteus maximus** – and both, along with some fat tissue, grant this area the thickness and the rounded contour that so many people are obsessed about.

- **Quadriceps** (tinted purple in the following figure): They are the biggest muscles in the human body, extending from the pelvic area all the way down to the knee. They lift the whole leg forward, as well as extend the lower leg. The most apparent are the **rectus femoris**, **vastus lateralis**, and **vastus medialis**, which make up most of the thigh's shape forming a large fleshy mass that covers the front and sides of the femur.

- **Calf** (tinted green in the following figure): The calf is composed of one muscle, the **gastrocnemius**, which is located at the back of the lower leg and divided into two segments, although this division is not very apparent. They extend from the knee to the heel and are responsible for extending our feet and assisting the walking movement. For that reason, this muscle tends to be pronounced in most people. Attached to it is the **calcaneal** tendon (also known as the **Achilles** tendon), the thickest in the human body, which also contributes to the shape of the heel.

- **Posterior thigh muscles** (tinted blue in the following figure): They are very big muscles as well, but are located at the back of the thigh. These muscles are responsible for flexing/bending the knee and also contribute to extending the whole leg backward. There are three muscles that are more pronounced in that area: the **biceps femoris**, **semitendinosus**, and **semimembranosus** muscles.

Additional muscles include the ones laterally attached to the **tibia** and **fibula** bones (not tinted), which make up the lower leg structure. Though they make a bit less of a difference when talking about the shape of the lower leg, they are responsible for tilting, flexing, and extending the feet.

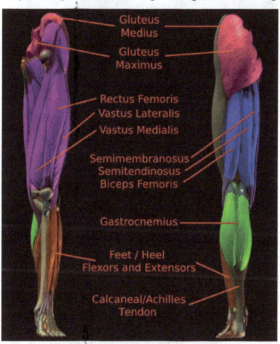

Figure 7.9 – Left leg muscle groups (model provided by the Z-Anatomy project at https://www.z-anatomy.com/)

Another part of the leg that makes a big difference is the knee. While it is small and mostly bone, it makes a significant difference in the shape of the middle part of the leg. This is why we can see it and distinguish most of it from the rest of the leg.

This is mainly because the kneecap sits extremely close to the skin, and is surrounded by tendons, muscle endings, and cartilage, which makes the knee important when working on a somewhat realistic depiction of a human leg.

The main structures that make a difference in the shape of the knee are the **kneecap/patella**, the **quadriceps** muscles (which go behind the kneecap and connect to the tibia, or **shin bone**), and the

medial and **lateral patellar retinacula**, which are branches of the tendon insertion of the quadriceps muscles and cross the knee horizontally (*Figure 7.10*):

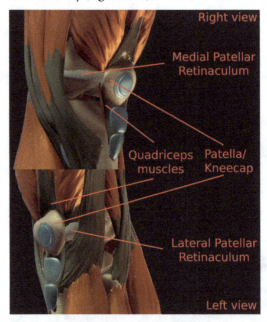

Figure 7.10 – Left knee (model provided by the Z-Anatomy project at https://www.z-anatomy.com/)

The kneecap becomes even more apparent when the leg is bent, while some of the surrounding structures may not appear so pronounced, depending on the person's weight and training.

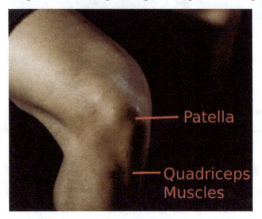

Figure 7.11 – Structures visible on a bent knee (image from https://www.pexels.com/)

In this case, the most noticeable structures are the patella and the segment of the quadriceps that appear below the knee.

Now, with all the main leg proportions and structures covered, let's have a look at the foot:

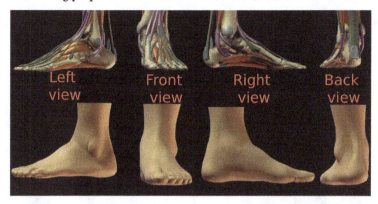

Figure 7.12 – Left foot (model provided by the Z-Anatomy project at https://www.z-anatomy.com/)

Foot size can vary a lot from person to person, but there are generally a few things to keep in mind, such as the fact that a foot is around the length of four big toes, and is often slightly bigger than one head, usually protruding forward a lot, with the tip of the big toe usually ending up at where the chest muscles end:

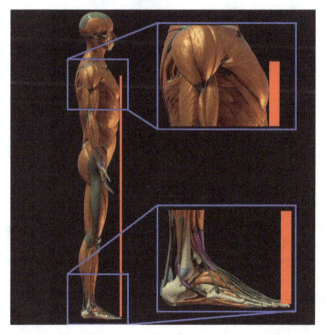

Figure 7.13 – Foot length in comparison to the body (model provided by the
Z-Anatomy project at https://www.z-anatomy.com/)

Notice as well that the heel doesn't go further than the calf muscles.

The toes also have an arch, which is formed by their tips and can help a lot when determining the size of each toe:

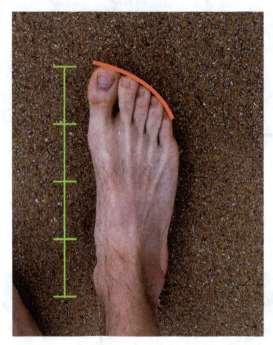

Figure 7.14 – Arch formed by the tips of the toes and big toe/foot proportion (image from https://www.pexels.com/)

The foot is composed mainly of small bones and tendons, having only a small number of muscles and fat tissue.

The main structures that shape the feet are as follows:

- **Bones** (tinted pink in the following figure): Most of the foot's shape is made out of bone, of which the largest is the calcaneus bone (or **heel bone**). Other bones that shape the foot significantly are the **talus** and the three **cuneiform** bones, located at the middle of the foot. The bones composing the toes are also significant to the foot's shape, as they grant another layer of stability when standing up.

- **Tendons** (tinted green in the following figure): The tendons are also noticeable in most people's feet, as they stay very close to the skin. These tendons are directly connected to some of the lower leg muscles and are most visible near the toes and the ankle. They assist the few muscles in the area in the flexion and extension of the toes.

- **Muscles** (tinted blue in the following figure): The main muscle groups influencing the shape of the feet are located on their sides and in the sole. There is one abductor muscle on each side and a larger flexor muscle in the middle of the sole.

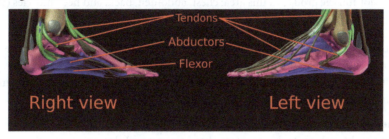

Figure 7.15 – Left foot structure (model provided by the Z-Anatomy project at https://www.z-anatomy. com/)

Now, with most of the lower limbs' proportions and structures covered, we'll have a look at the chest and abdomen, which keep everything together.

Chest and abdomen

For these parts, there's also a considerable amount of variation from person to person, depending on several factors such as diet and genetics, along with daily routines. But no matter the body type, the underlying structure remains the same:

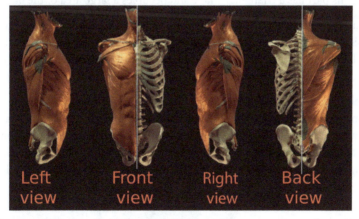

Figure 7.16 – Chest and abdomen (model provided by the Z-Anatomy project at https://www.z-anatomy. com/)

On a grown adult, the chest usually is three heads tall and three heads wide at the shoulders (if the heads are put side to side, vertically), and it tends to get thinner near the waist, where there are no bones. Again, this aspect also depends on body type.

Also, note that the spine is curved and doesn't go straight down. That's a mistake many do when starting. Rather, it has a slight "S" shape, starting at the neck.

There are four major muscle groups in this area: chest, abdomen, back, and neck muscles (which may not be a part of the chest itself, but has muscles connected to it):

- **Chest** (tinted pink in the following figure): The chest is mainly composed of one thick fan-shaped muscle that makes up the bulk of the chest and lies under the breast, called the **pectoralis major**. It extends from the middle of the ribcage to the armpit and plays a fundamental role in pushing the arms forward.

- **Abdomen** (tinted green in the following figure): The abdomen is composed of two muscles, both playing a big role in its shape: **external** and **transversal** abdominal muscles. And although both of them contribute to the shape, only the external muscle is visible from the outside. They are responsible for curving the chest forward and, when properly trained, can get very pronounced in the body.

- **Back** (tinted blue in the following figure): The back muscles are also very big, and there are two of them that make the most difference to the shape: the **trapezius** and **latissimus dorsi** muscles. The trapezius is composed of three parts. The first is the **ascending** part, which medially rotates and depresses the **scapula** (or **shoulder blade**) and extends from there until the lower back of the head. Next is the **transverse** part, which retracts the scapula and is located in the upper back, extending from the spine to the shoulder and attached to the end of the scapula. And finally, we have the **descending** part, which is the biggest of them all and supports the weight of the arm. This part has a triangular shape and occupies a large area of the back.

  The latissimus dorsi muscle is the largest muscle in the upper body, occupying most of the back and extending to the sides until under the arms, and it is responsible for most of the movements of the shoulder joint. As this muscle is fixed to the spine, it also has a role in the extension and lateral flexion of the lumbar spine.

- **Neck** (tinted purple in the following figure): There are two apparent muscles in the neck. The **platysma** muscle superficially covers the front portion of the neck. When contracted, it causes slight wrinkling and covers the **sternocleidomastoid** muscle, which extends from the **clavicle** all the way to the back part of the head. The primary actions of this muscle are flexing the neck and rotating the head to the opposite side. This muscle is more apparent than the platysma.

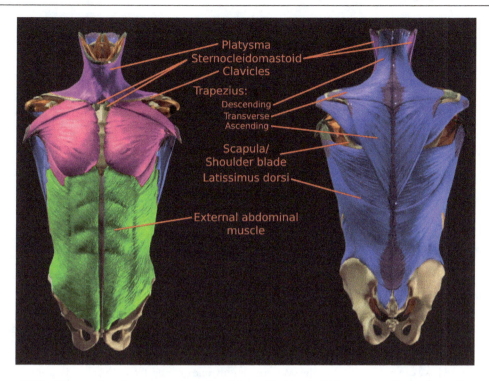

Figure 7.17 – Chest and abdomen structures (model provided by the Z-Anatomy project at https://www.z-anatomy.com/)

It's worth noting that humans have bilateral symmetry, meaning that the right side is symmetrical with the left, and this is no exception.

Now, let's have a look at one of the most important and complex parts of the human body: the face.

Face

Now, the face is one of the most defining features of the majority of characters, and it's extremely important to get it right for the character to look believable. We can notice when something is wrong with a human face with incredible ease, so any minor mistake can end up making our character look weird, wrong, or, in some cases, uncanny.

For that reason, we'll mostly look at real people instead of anatomical models and analyze the main proportions of the facial features.

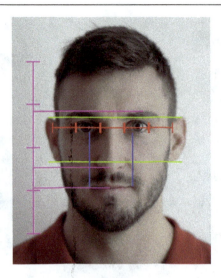

Figure 7.18 – Human facial proportions (image from https://www.pexels.com/)

Several markings of different colors have been made in the preceding figure to give us a better idea of where the facial features are located relative to each other, as well as relative to the head in general.

Marked in red, we can see two things: the eyes are roughly one eye apart, and the skull is around five eyes wide. This is not to be confused with the eye sockets though, which are bigger, starting at the edges of the nose bridge and extending around the edge of the skull:

Figure 7.19 – Human skull (image from https://www.pexels.com/)

Moving on, marked in blue, we can see the size of the mouth relative to the eyes, with the edges of the lips reaching about a third of the length of each eye when relaxed. Sometimes, they can reach the middle of the eye as well.

Marked in green, there's the general size of an ear. The top roughly aligns with the eyebrows and the bottom usually ends just above the jaw.

Finally, if we divide the face into four equal segments, which are marked in pink, we notice that the first and last quarters are composed only of the forehead and chin, respectively, and all the facial features are usually located in the two middle quarters of the face:

- The top of the eyebrows is just below the end of the first quarter

- The middle of the eyes roughly lines up with the middle of the second quarter

- The nose bridge starts at the eye level

- The tip of the nose is located roughly at the center of the third quarter

- The mouth is right above the top of the fourth quarter, with the top lip being in the third one and the bottom lip, in the last

Note that the location and proportion of the facial features can vary depending on several factors, so while sculpting, we do have a bit of wiggle room without it looking very weird, even while working with ideal proportions as we are doing.

A good tip for the face size is that we are capable of covering most of our facial features with one hand, so that's a good technique to know whether the sculpted hand or head is too big or small.

Note that some of the parts of the face may not follow a general rule like the ones mentioned here, such as the nose and cheeks. These can vary a lot from person to person as there are several types of noses, and depending on genetics and diet, the cheek may have a larger or smaller amount of fat, which can make the cheek area protrude inward (making the cheekbones more pronounced), outward, or be flatter.

Now, with all the human body's general proportions and shapes analyzed, let's practice this by sculpting a base mesh with realistic proportions for a character.

Remember that this is a complex task and should ideally be approached when you are extremely comfortable with the sculpting tools available.

Blocking out the base mesh

Reference, reference, reference. This is the most important word of this stage because we might know how the human body is shaped and its bigger and most defining features, but even then, sculpting humans and most creatures purely from imagination is generally a bad idea, especially as a beginner. Even the most experienced artists out there use references for their work. So, before putting our hands to work, it's extremely important to have a good set of references for what we're going for. Remember, though, that humans are extremely complex to get right, so don't feel discouraged if, at first, your model looks wrong. We'll try to simplify the process as much as possible.

For our case, we'll be making both male and female base meshes using mostly ideal proportions.

Let's make this first sculpting iteration with as little detail as possible to keep things clean, as we're just trying to get the core shapes and silhouette down. After that, we can go back and refine the sculpture to add more detail. The strategy now is to visualize the bigger shapes and simplify the anatomy as much as possible.

We'll start with the head, and it may sound counter-intuitive, but it will be better to add a cube and subdivide it using the **Multiresolution** modifier, which will initially be set to **2** subdivisions. This will turn the cube into a very low-resolution sphere:

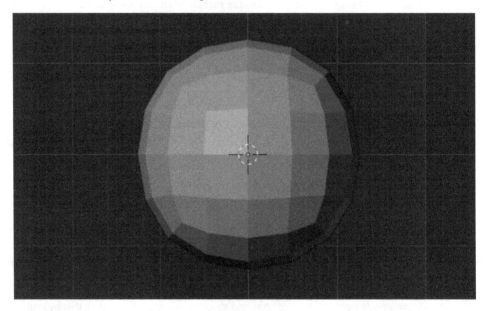

Figure 7.20 – Subdivided cube

Now, in **Sculpt** mode, we'll begin to shape the chin and the jawline by using the **Grab** brush at a decent size to pull the lower vertices at the center downward. Make sure that **Symmetry** (on the **X** axis, in this case) is enabled to avoid bigger issues later on.

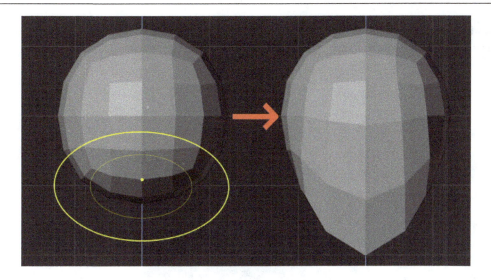

Figure 7.21 – Initial shaping of the head

We can now proceed to flatten the sides of the head as our skull gets flatter on the sides, still using the **Grab** brush:

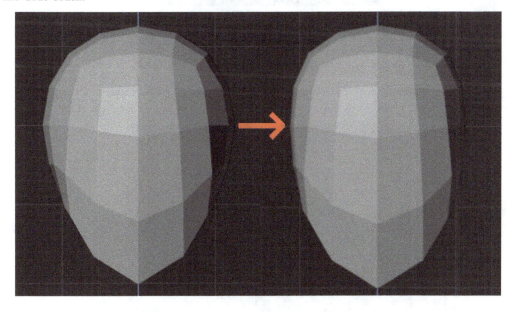

Figure 7.22 – Flattening the sides

Now, with the general shape of the head, we can use the **Array** modifier set to **8** copies in the **Z** axis to measure the height of our character, indicating where several features should be located:

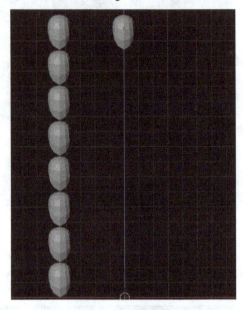

Figure 7.23 – Measuring the height of the base mesh using the head

Now, back to the head; we'll add another level of subdivision and better shape the chin and jawline.

Keep in mind that, generally, a rounder, thinner jawline with a sharper chin will look more feminine than a wider and sharper jawline. It's also useful to start shaping the jawline from the sides:

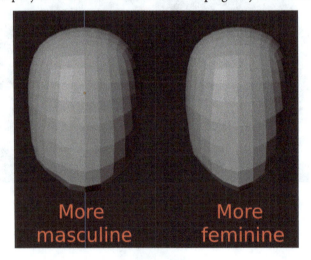

Figure 7.24 – Masculine and feminine jawlines

Now, we'll proceed to add the neck, using another cube with the **Multiresolution** modifier, and pull it down while also going slightly backward, as our spine doesn't go straight down:

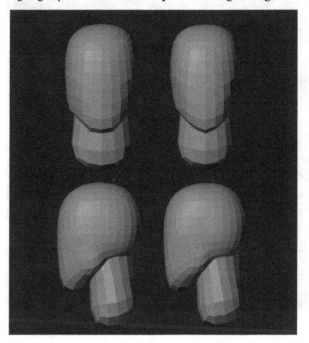

Figure 7.25 – Neck shaped using a subdivided cube

Now, with the neck's general shape in place, we can move on to the main body, again, keeping it as low poly as possible and with symmetry enabled. We'll separate the area into three individual parts: chest, abdomen, and waist.

Starting with the chest, it starts right at the base of the neck and extends down until just below the second head measuring the height. Remember though, that men usually have wider chests and shoulders than women, and that in both cases, the shoulders are wider than the base of the chest.

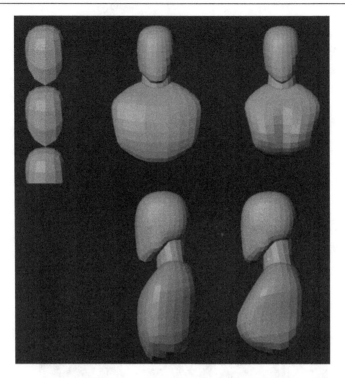

Figure 7.26 – Male and female chests added

Note that the spine is not going straight down but rather at this portion, is protruding slightly backward at the upper portion and slightly inward in the lower part. Also, notice that the female chest goes forward a lot more due to the breasts.

Next, the abdomen. We'll use the same technique used until now for all the parts: start with a cube and subdivide it using the **Multiresolution** modifier, then shape it according to the part we're shaping using mainly the **Grab** brush.

The abdomen is also usually wider at the top, in females and males, although it tends to be slightly thinner in women both in wideness and depth. The abdomen has a trapezoid shape and extends down until just under the third head.

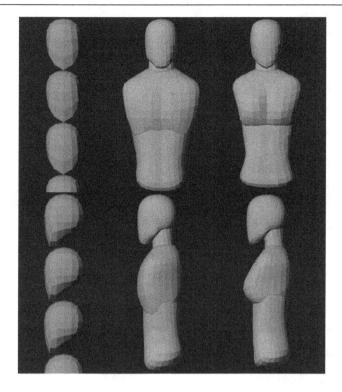

Figure 7.27 – Male and female abdomens added

At this stage, the middle of the abdomen and chest were pushed inward to more clearly define the division between the muscles on the right and left sides. This was also done in the back of each base mesh.

Feel free to go back and adjust the different parts to better fit each other, adjust/fix proportions, and so on. You can quickly switch between the parts in **Sculpt** mode using *Alt + Q* while hovering your cursor over the mesh you want to sculpt. It's much easier to do these major shape changes now than when we add more detail to it later.

Now, let's move on to the waist. This uses the same technique as before: start with a subdivided cube set to **3** subdivisions on the **Multiresolution** modifier and shape it. The shape of this area is heavily influenced by the pelvis and extends until around the bottom of the fourth head measuring the height.

We'll shape it leaving space for the upper legs, so this means that the pelvic area will have a shape that resembles a diaper. Also, remember that women usually have wider hips than men so that the former can give birth.

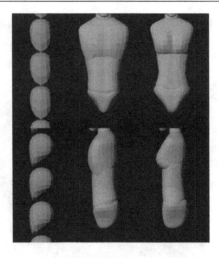

Figure 7.28 – Male and female waists added

We won't make the gluteus now. Instead, we'll first make the legs. This way, we'll have more information to work with and determine how the gluteus should look in each one of the base meshes.

We'll separate the legs of each of them into three parts: upper leg, knee, and lower leg.

The upper leg extends until around the bottom of the sixth head, and in its upper portion, it's usually as thick as the waist and almost as wide as one head, getting slightly smaller as the leg goes down. Again, women's thighs tend to be thicker than men's.

We'll do the same thing as before, except this time, we'll add a **Mirror** modifier to the leg and disable symmetry, as we have two of them and they are not symmetrical themselves, with **Mirror Object** set as the waist.

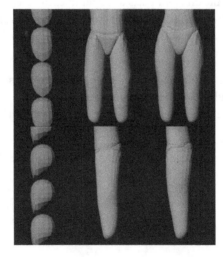

Figure 7.29 – Male and female upper legs added

See how the thigh muscles protrude outward in the upper portion and slightly inward in the lower part of the thigh.

Now, for the lower leg, we'll just duplicate the upper leg with *Shift + D* and reshape it to the shape we're going for. We can also shape another subdivided cube for the knee into an upside-down triangular shape while we're at it, as it's a small part with a simpler shape.

Again, don't forget your references!

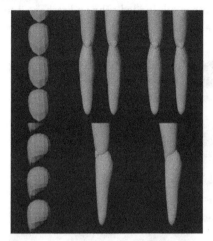

Figure 7.30 – Male and female lower legs and knees added

Now, with the legs done, we can go back and add the gluteus to our meshes. We'll shape it similarly to a water balloon for both of them, as the fat in this area accumulates in the bottom. This prevents it from having a perfectly spherical shape, which would look unnatural.

Remember that the gluteus is usually wider and rounder in women than in men.

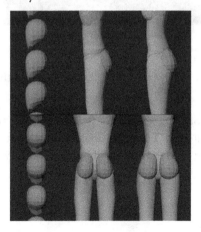

Figure 7.31 – Male and female gluteus added

As you can see, each side of the gluteus is a bit far away from the other, but this is intentional, as in the future, when we proceed to merge and remesh everything, there may be a loss of detail, merging what should be two separate sides if we left them too close to the line of symmetry.

Now, we'll move on to the arms. And we'll divide them into four parts: shoulder, arm, forearm, and elbow. If you want, you can also separate the arm into biceps and triceps for a bit more detail in this early sculpting stage.

For the shoulder, we can add another subdivided cube and reshape it to more of an elliptical shape. We'll make the arms at around a 45° angle in order to easily pose them, which will be covered in a future chapter.

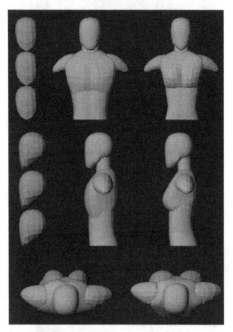

Figure 7.32 – Shoulders added

As you can see, there's not much difference in the actual shape of the shoulders between males and females.

For the arm, we'll shape it using one cube instead of separating it into the biceps and triceps muscle groups. The arm muscles are rotated 90° from the shoulder muscles, starting on the sides and under the deltoid muscle, and extending until the bottom of the third head (when put straight down), near the belly button.

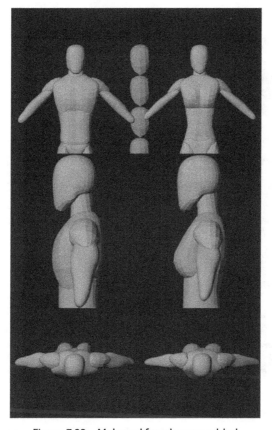

Figure 7.33 – Male and female arms added

Notice that female arms tend to be a bit thinner. Also, see how the triceps protrude outward at a higher point in the arm than the biceps, but the biceps are longer and rounder than the triceps, usually extending until near the elbow.

The forearm looks similar to the arm, but its muscles are rotated 90° relatively to the arm, just like with the arm and the shoulder.

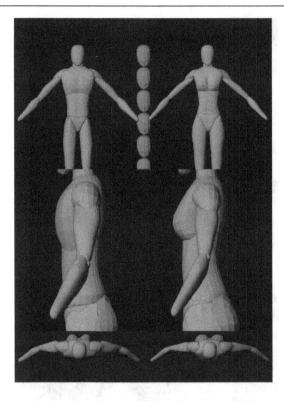

Figure 7.34 – Male and female forearms added

Notice how the brachioradialis muscle protrudes outward at a slightly higher point than the flexors, which are rounder as well.

Remember that at any point in this early block-out stage, making changes to the proportions or anatomy is completely normal, as we tend to notice more "flaws" with the model we're making as we add more details to it and have more information to work with.

Now, the only things left before we move on to refining the sculpt and adding more detail are the hands and feet, which are among the more complex things to make.

We'll start with the feet since, although they are harder, they are still easier than the hands.

Looking at it from the sides, a foot is quite similar to a trapezoid shape:

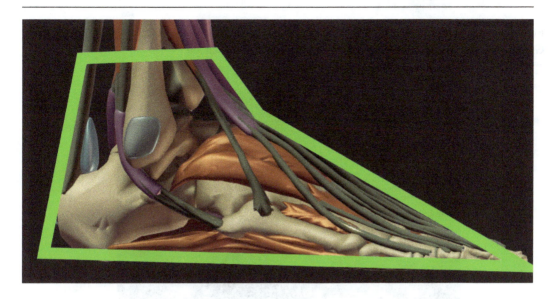

Figure 7.35 – Trapezoid shape of the foot (model provided by the Z-Anatomy project at https://www.z-anatomy.com/)

We'll replicate that shape, still using mostly the **Grab** brush:

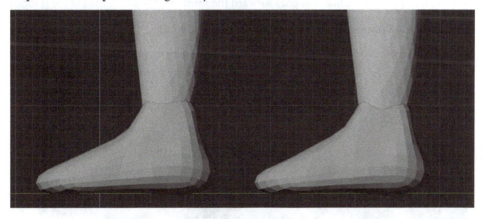

Figure 7.36 – Male (left) and female (right) feet side view

As you can see, male and female feet have a very similar general shape, although women tend to have thinner and smaller feet.

Remember to keep a mostly flat bottom whenever you're doing any sort of foot/footwear.

Before moving on to the toes, it's important to also make an arch on the front of the foot using the **Grab** brush, just like when we covered the foot anatomy in this chapter:

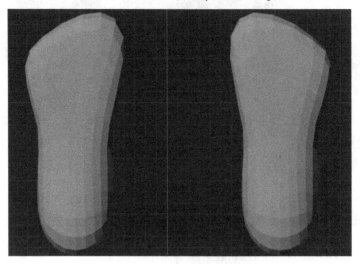

Figure 7.37 – Arch made in the foot (seen from below)

Notice how the inner part is the one that goes forward the most. This is where the big toe is located.

The toes themselves are simpler, mainly resembling medicine capsules:

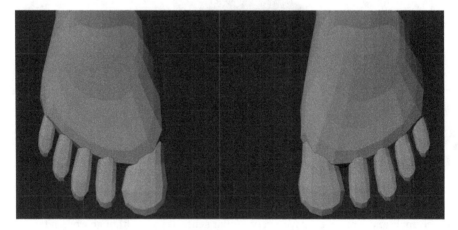

Figure 7.38 – Toes added (seen from above)

See how they have spaces between each other; this is for the same reason as we did it for the gluteus. Notice how the toes occupy the entirety of the foot's front and that the next toe is never bigger than the previous one, starting from the big toe.

Also remember the big toe/foot ratio of 4:1, which is a good way to determine the size of the big toe relative to the foot.

Now, let's move on to the hands, which are more complex. For that reason, we recommend working on the hand separately, keeping it standing straight up while working on it and only positioning it when we're finished with the main shape.

We'll separate the hand into three parts: palm, fingers, and thumb. Reference is very important here as well.

For the palm, we'll shape a subdivided cube into a rounded tall trapezoid with an arch on top:

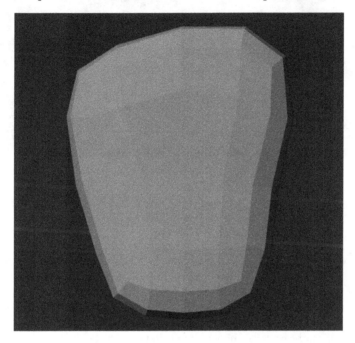

Figure 7.39 – Palm base shape

This was made using a subdivision level of **2** on the **Multiresolution** modifier, initially.

Onto the fingers, we'll use a technique similar to the toes, shaping a cube into a curved sausage-like shape for the middle finger, as it's the biggest and has an almost 1:1 ratio with the palm, and then duplicating this finger for all of the other fingers of the hand, keeping the arch between their tips in mind while doing so.

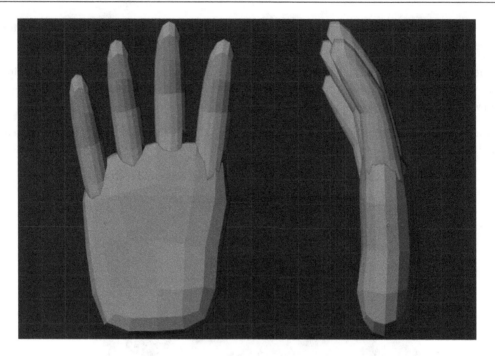

Figure 7.40 – Fingers added

Notice that the fingers are slightly more spaced than they should be. That's for the same reason the toes are also spaced. We can fix that once we're done with our sculpting.

Also, see how there's little to no change in width and wideness on all of the fingers. That's a very common mistake, where people make the bottom of the fingers much wider and/or thicker than the top.

For the thumb, we'll duplicate the index finger and put it in place, at around a 45° angle from the palm, near the bottom, rotating the finger so that the flat tip of the finger is pointing inward instead of to the front, then inflate the bottom a bit with the **Inflate** brush at a decent size:

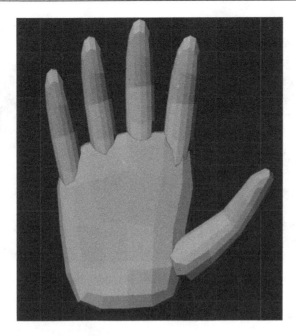

Figure 7.41 – Thumb added

Now, let's not forget the muscle group near the thumb, which is very apparent. For that, we'll duplicate the palm and smooth it out to an oval shape using the **Smooth** brush, then back to the **Grab** brush, and shape it like a big bean that surrounds the bottom right palm and the base of the thumb:

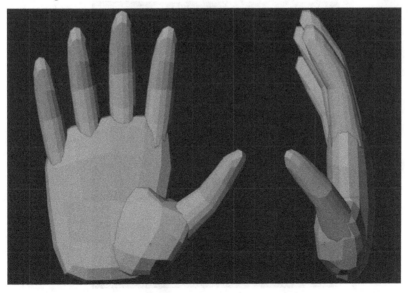

Figure 7.42 – Thumb muscles added

Now, with our hand looking decent, we can position and scale it on our base meshes, with the palm facing down and the thumb pointing forward:

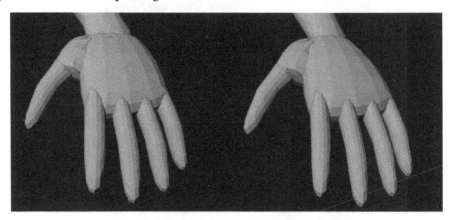

Figure 7.43 – Hands positioned on the male (left) and female (right) base meshes

The fingers on the female base mesh are slightly thinner, as well as the palm.

When positioning the hands, don't forget to adapt the shape of the bottom part of the forearm to better resemble a wrist and fit the hand better.

Now, we're finished with the block out of our base meshes, and here is a full body image of what they look like:

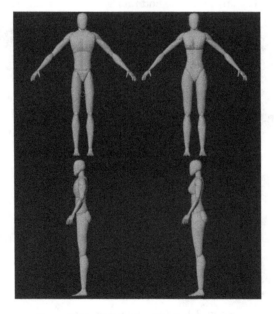

Figure 7.44 – Full body of the base meshes

Perfect, they look decent. Now, it's a good time to go back and check that you're happy with all of the proportions because, from the next steps on, it will become increasingly difficult to implement any major changes.

We highly recommend saving your progress and creating a backup copy of the models in case anything goes wrong from here.

Summary

In this chapter, we covered the main anatomical elements that make up a human when sculpting, such as bones, muscle groups, and tendons, going deeper into how each major muscle works and how it affects the shape of the human body, as well as the main differences between male and female bodies. We also covered how to block out male and female base meshes with mostly ideal proportions, simplifying the anatomy as much as possible by using simple shapes to convey the core shapes of a human body to build a good-looking silhouette.

At this point, you are more knowledgeable when working with anatomy and can construct a simple block out of the core shapes that compose the human body.

In the next chapter, we'll refine the sculpture we have now, applying more of the anatomy seen in this chapter.

Refining the Base Meshes

Now with the block-out finished, we will begin to add more detail to the sculpting by going back and adding more geometry to each part with the necessary details, going into further depth on the distinct aspects of the human anatomy covered in the previous chapter. This process is essential in order to properly refine the look of any character since here is where we make our character believable and give it some personality.

In this chapter, we'll cover the following:

- How to refine the block out, adding more detail to each part
- The suggested brushes each step of the way based on the shape of the area
- How to properly join the meshes, smoothing the division between each part of the meshes

By the end of this chapter, you should be able to sculpt decent-looking humanoid figures.

Refining the head and face

Starting with the head and face, we can start to sculpt the main facial features and muscles by either applying the **Multiresolution** modifier and re-meshing the head using *Shift + R* then *Ctrl + R* or keeping the multiresolution modifier as it is and increasing the subdivisions. The latter method offers the possibility of going back and forth between levels of detail, which can be useful. If you decide to keep the modifier though, it's good practice to click **Apply Base** (located under the **Shape** menu on the **Multiresolution** modifier) every time you make a bigger change to the shape of the mesh.

Next, we smoothed everything out using the **Smooth** brush and traced a line across the face at around a third of its height using the **Draw Sharp** brush, marking where the eyes should be:

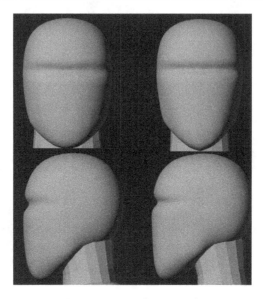

Figure 8.1 – Eye line carved

Now, we'll carve the holes for the eye sockets – not the eyes, the eye sockets. They are very large within the skull, as covered earlier, so don't be afraid to make them bigger, and remember that *Ctrl + Z* is your best friend.

The eye sockets were carved using the **Draw** brush with symmetry enabled:

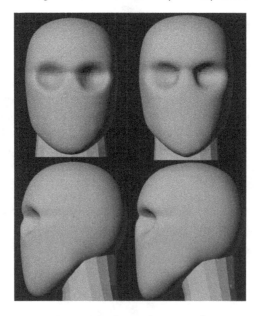

Figure 8.2 – Eye sockets carved

Now, using the **Scrape** and **Clay Strips** brushes, we'll flatten the area underneath the eyes and where the cheeks are since there's no bone there to make it as round as it is at the moment.

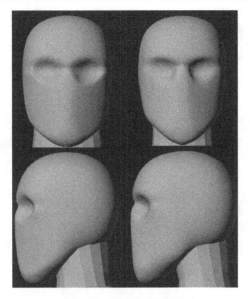

Figure 8.3 – Cheeks flattened

Now, for the nose – first, we'll push the geometry of the area outward using the **Clay Strips** brush to define the general size of the nose. Apply it until you have something protruding out of the mesh:

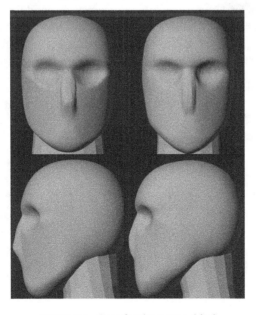

Figure 8.4 – Base for the nose added

Now that we have something sticking out, we can use **Grab Brush** once again to define the profile of the nose. Finding a reference would be extremely useful here since there's a wide variety of nose profiles. In our case, we'll make a nose with a flatter profile in both cases.

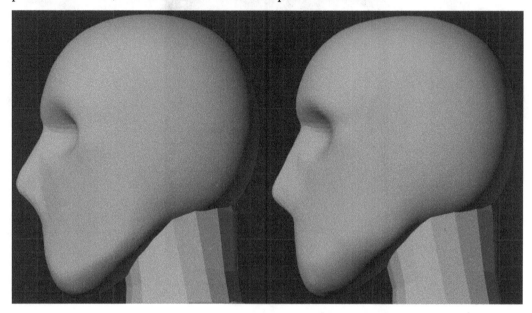

Figure 8.5 – Nose profile

Now we'll make the nostrils. First, carve a bit into the geometry around the nose in order to make it slightly flatter, and then we'll proceed to inflate the nostrils using the **Inflate** or **Clay Strips** brushes. Don't be afraid of making the nostrils too big, as you can shape them using the **Grab** brush afterward. Make sure that you inspect the area from all angles to ensure that it doesn't look strange from any point of view.

We can also carve the nostril holes when we're finished and define them more using the **Draw Sharp** brush around the nostrils by meshing it again at a higher resolution or subdividing the mesh:

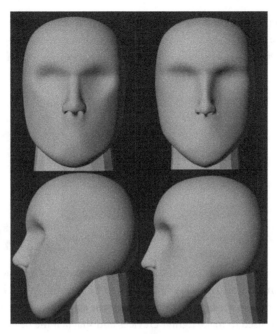

Figure 8.6 – Nose finished

We'll move on from the nose for now, but it's always possible to come back and tweak the shape according to your desired design.

Now, we'll add the ears, which are currently missing. We'll start by adding a subdivided cube, and making it taller, slightly oval, and flatter:

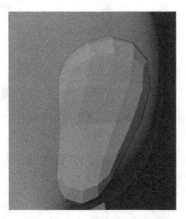

Figure 8.7 – The base shape for the ear

Now, we'll increase the resolution, smooth it out with the **Smooth** brush, and start shaping the interior to more of what's inside a human ear. Use a reference for this.

We'll still keep it fairly simple though, as we don't need a 100% accurate shape since the ear is a small part and this is only a base mesh:

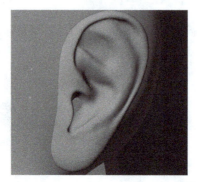

Figure 8.8 – Interior of the ear shaped

It's not exactly perfect, but it serves our purposes, and we can always come back and tweak it to our liking. This was made using mainly the **Clay Strips** brush for carving in the hole for the ear canal, combined with the **Draw** brush to shape the cartilage inside the ear (which resembles a curved 'Y'). The **Draw Sharp** brush was also used to make tighter crevices and sharpen the peaks around the ear. If you notice some weird bumpiness when dealing with higher resolution meshes, it's a good idea to use the **Scrape** or **Flatten** brushes combined with the **Smooth** brush to get the surface looking smooth again.

Now with the ears done, we can go back and adjust the jawline with the **Grab** brush to push it down since it's a bit too steep right now. We can use the ears as a point of reference since they are located roughly at the middle of the sides of the skull and extend down to just above the jawline:

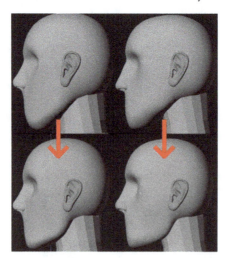

Figure 8.9 – Jawline fix

Notice how the ear has a slight tilt; this is also important to make the face believable.

Moving on, we have the eyes. It's extremely important to get them right, as they are, according to a very smart person, "the window to the soul" and we tend to notice quickly when there's something wrong with them in a drawing, or in a sculpture, in our case.

We'll start by adding a UV sphere for the eyeballs so that we have something to cover the eyelids with. Just like that, add the sphere, position it, and scale it on the face:

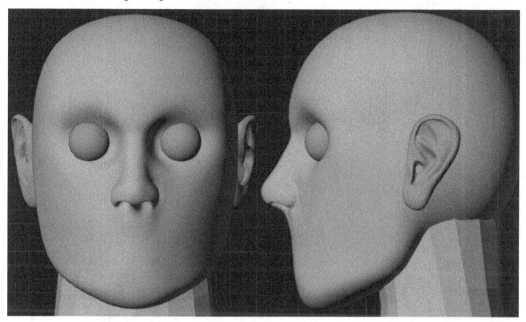

Figure 8.10 – Placeholder eyes placed

If you're using a real-world scale, an adult's eyeballs are usually 21 to 27 millimeters (about 1.06 inches) in diameter.

We'll start by just setting the eyelids over the eye using the **Draw** brush to bring the eyelids over the eyes and roughly shape them:

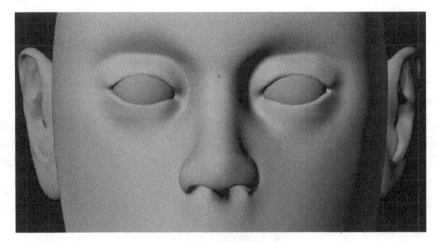

Figure 8.11 – Rough eye shape done

Now we can start shaping the eyelids however we want. There are several shapes for the eyes that you could use, but in this case, we'll opt for an almond-shaped eye, so make the upper portion curvier and the lower eyelid flatter:

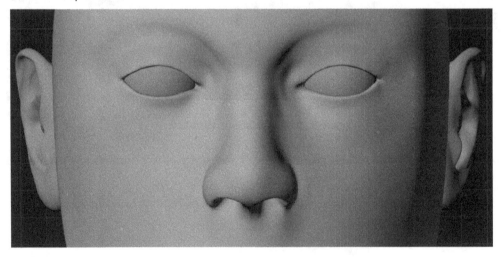

Figure 8.12 – Eye shape refined

The shape was refined using a combination of the **Grab**, **Draw**, **Clay Strips**, and **Scrape** brushes for the shape and the **Draw Sharp** brush for sharpening the edges of the eyelids, as well as the **Smooth** brush for the hard edges.

Don't forget that eyelids are quite thick, both the upper and lower lids. We can also sharpen the edges of the thickened lids using the **Draw Sharp** brush:

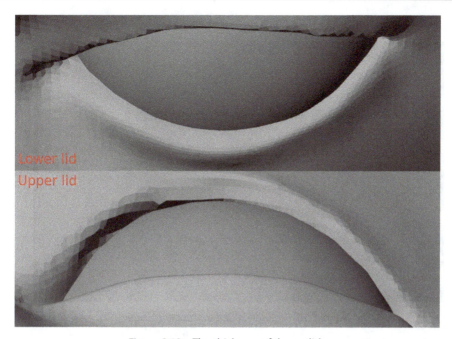

Figure 8.13 – The thickness of the eyelids

Lastly, we can adjust the bones above the eye in case it looks out of place now that we have the eyes. Again, using the **Draw** brush to make them more apparent, forming a hood-like shape above the eyes:

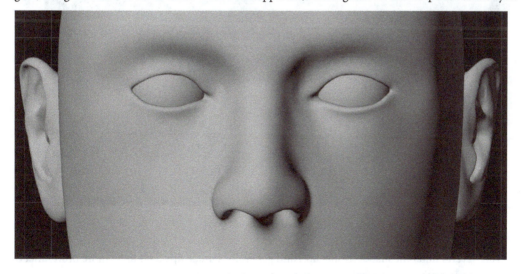

Figure 8.14 – The bone above the eyes shaped

Good – let's move on for now.

For the mouth, we'll start by using the **Clay Strips** brush to add some volume to the area:

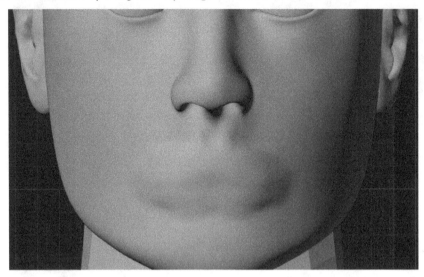

Figure 8.15 – Volume added to the mouth area

Now, using the **Draw Sharp** and **Grab** brushes, we can start tracing the shape of the lips. Don't forget to use a reference!

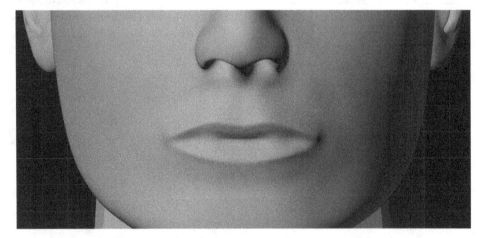

Figure 8.16 – Lips traced

There are several different shapes for the lips, and generally, female lips tend to be more pronounced and rounder. In both cases, though, the upper lip usually protrudes out more.

Now, with this base, we can start puffing up the lips, fixing them from all angles, and tweaking their shape using some of the rounder brushes available, such as the **Draw** and **Inflate** brushes:

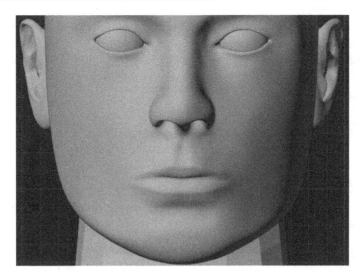

Figure 8.17 – Lips further shaped

Now, with all the facial features added, it's a good time to look back and tweak anything that might have been looked over or that doesn't look right now that we have added all the features. Now is the time for smaller tweaks to the shape of the skull and/or the facial features, since we can now look at the face as a whole.

Here's the final result for the male and female faces:

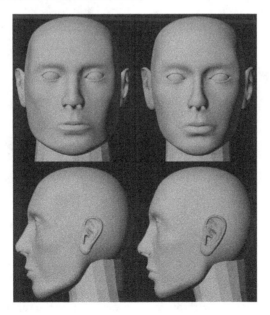

Figure 8.18 – Final result for the male (left) and female (right) faces

As you can see, the cheekbones were made stronger (using the **Clay Strips** and **Smooth** brushes) and the sides of both heads were flattened slightly more. Remember that we're trying to make the base meshes with as little "personality" as possible, let's say, along with a reduced amount of detail. This is so that we can more easily tweak the features later when we're making actual characters to convey specific personalities.

Now with the face done, we can finally move on. To the neck, we go!

The neck

The first thing we'll do is add more resolution to it, either by adding more subdivisions to the **Multiresolution** modifier or by applying the modifier and re-meshing it at a higher resolution so that we can shape it better. We kept the modifier and set it to the fifth level of subdivision.

Let's start with the **sternocleidomastoid** muscles, which extend from the back of the head to the clavicle. They are less visible on the sides of the neck when relaxed but become more visible as they get close to the front. These muscles can be more or less apparent depending on how thin and/or athletic the person is. We are making both base meshes in relatively good shape so they will be slightly more apparent.

Using the **Draw** and **Grab** brushes, we can start adding volume to the area, roughly shaping where the muscle should be. Remember to keep symmetry enabled.

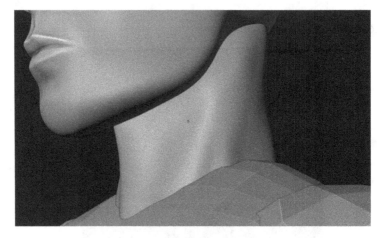

Figure 8.19 – The Sternocleidomastoid muscle added

Notice how the muscle makes it so that the neck curves inward at the front. That's because there aren't many bigger structures there, only smaller muscles. This was made using the **Grab** and **Draw** brushes as well.

We can now push the frontal middle part forward in order to shape the cartilage that is more visible in the throat area. Make sure to push the upper portion slightly more:

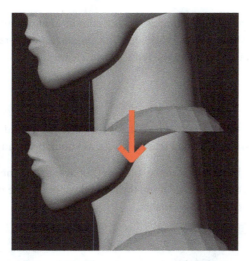

Figure 8.20 – Throat area pushed forwards

We're done with the front part for now, and we can start shaping the back of the neck, mainly the **descending trapezius**, which starts at the back of the cranium and extends to near the shoulders. Many people get this muscle wrong, so pay extra attention to the reference.

The Descending Trapezius has a triangular shape with one point on the back of the head, one at the shoulders, and one in the middle of the upper back, near the spine. It's common to have a slight crease in the upper back due to the muscles.

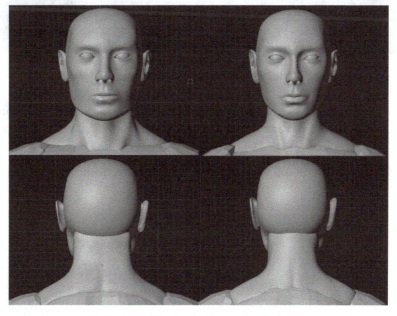

Figure 8.21 – The descending trapezius added

Notice how in both the male and female models, it has a contour that resembles a steep ramp, which gradually gets flatter as it gets closer to the shoulder. The female trapezius tends to be steeper in the upper portion and less pronounced due to the reduced muscle mass, generally.

The chest

For the chest, we're doing to use a different approach for each one of the models since there's generally a big difference in shape between male and female chests, as you may already be aware.

For the male chest, we'll start by adding more resolution (not as much as the previous ones, since it's a slightly flatter area with less detail) and defining the general volume of the pectoral muscles and the clavicles using the **Clay Strips** and **Draw** brushes.

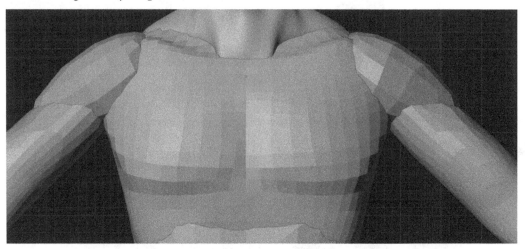

Figure 8.22 – Male chest muscles and clavicles roughly defined

Notice how the pectoral muscles have a generally square shape with a rounder outer contour and extend until under the armpit. The clavicles tend to be angled upward slightly as well for both males and females.

Now we can add more resolution and refine the shape more to better define both features:

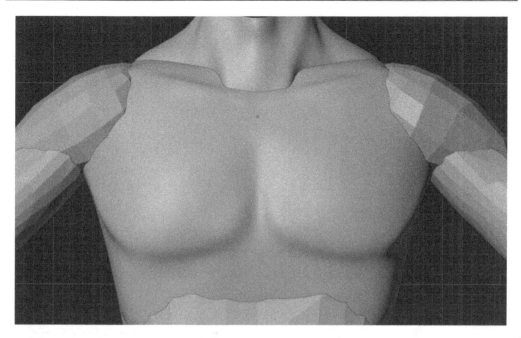

Figure 8.23 – Male chest refined

The edges of the pectoral muscle and the crease under it were sharpened using the **Draw Sharp** brush and the **Grab** brush was used to drag the sides to under the armpit area:

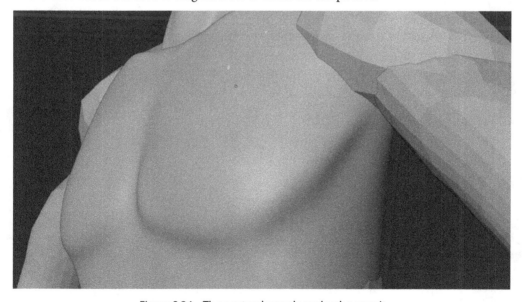

Figure 8.24 – The pectoral muscle under the armpit

We can also shape the upper part of the abdominal muscles using the **Clay Strips** brush to add some volume to the area, following the shape of the pectoral muscles until we reach the sides:

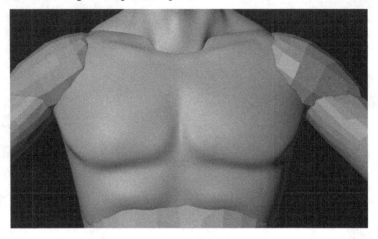

Figure 8.25 – Upper abdomen muscles shaped

As you can see, the upper part of the abdomen was puffed up a bit. It's subtle now but will become more apparent when we work on the abdomen.

Now let's move to the female chest before we make the upper back of both, which share a lot more similarities.

As with the male chest, we'll start by adding more resolution to shape the clavicles and better adjust the shape of the breasts, this time using the **Grab** and **Inflate** brushes instead of the **Clay Strips** brush to create more roundness.

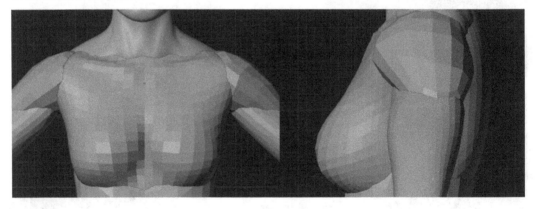

Figure 8.26 – Female chest better shaped

In this case, the gap between the breasts was tightened and the tips of the breasts were pulled apart slightly more. The clavicles were also added, and keep in mind the clavicles of females tend to be more pronounced.

Now, we can add more resolution to the mesh and refine the shape more, shaping the clavicles and breasts better and adding the slightly visible part of the pectoral muscle at the sides, going under the armpit.

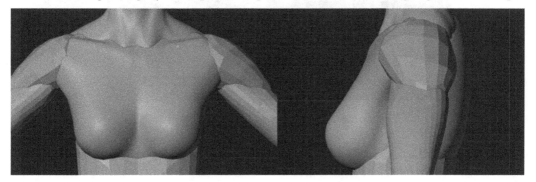

Figure 8.27 – Clavicles and breasts refined

You can use the **Pinch** brush to tighten the gap between the breasts and the **Draw Sharp** brush in case you want to sharpen something, such as the edges of the pectoral muscles.

Moving on, we'll make the upper back of both characters.

The upper back

The male and female upper backs thankfully share more similarities than the front of the chest.

Mainly, we have the **scapula** (or shoulder blade) and the upper portion of the **ascending trapezius** to worry about. We can also add a few smaller muscles, depending on how muscular we want our base mesh to be, but let's focus on the bigger things first since they make more difference.

We're setting up the base meshes to be in good physical condition, so the muscles will be more apparent. Let's start with the **transverse trapezius** since it will be a good reference for the other back muscles.

Repeat after me: don't forget your references!

The **transverse trapezius** starts at the shoulders and extends to the spine, forming a rectangular shape:

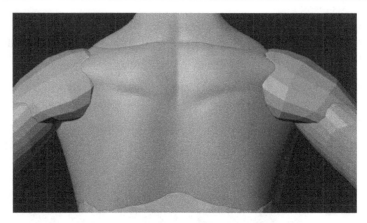

Figure 8.28 – The Transverse Trapezius shaped

For these muscles, it's much easier if we use the **Clay Strips** brush to build up volume and the **Smooth** brush afterward to make the surface smooth again since this brush leaves harsher edges. If you notice you added too much volume, you can use the **Grab** brush to bring it back to a point where it looks decent.

Now let's make the **ascending trapezius**, which extends to the middle of the back, but the upper portion is more apparent. It also extends from the shoulders to the middle of the back while forming a triangular shape, getting thinner as it goes down:

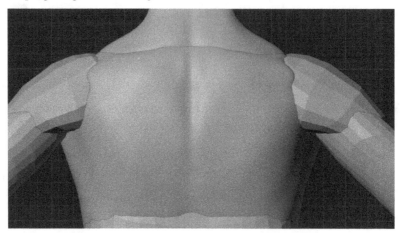

Figure 8.29 – The Ascending Trapezius shaped

Remember that the **transversal trapezius** we shaped before will become less apparent.

While we're at it, we can also shape the upper portion of the **latissimus dorsi** muscle, commonly referred to as the "lats." It's way more apparent in bodybuilders but it can affect the general shape of the back in less athletic people as well.

It extends from under the armpit to the lower back, and it also wraps around the ribcage.

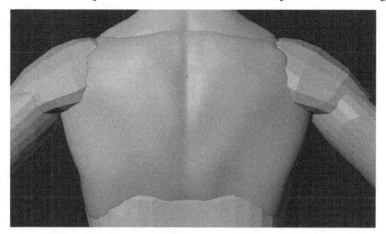

Figure 8.30 – The upper part of the latissimus dorsi muscle added

Notice how, by adding the lats, the trapezius became slightly more apparent. You can accentuate this feature using the **Clay Strips** brush to carve inward around the trapezius.

The female upper back has the same structures, but the muscles tend to be less apparent, again, due to a lower muscle mass. This causes the shoulder blades to become more defined, which also happens for thinner people in general.

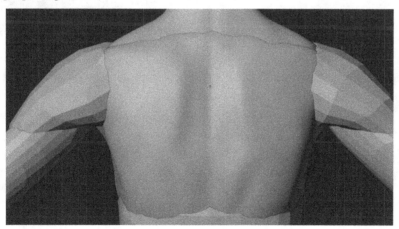

Figure 8.31 – Female upper back

Notice how everything is smoother and the shoulder blades are more visible. The lats are also less apparent.

Let's move on from the torso for now and start working on the arms.

The arms

The arms can be a little trickier to work with since there's a lot of variety depending on how athletic the person is. For that reason, references are especially useful here.

Let's start with the shoulders. They are simple to make, as their shape still is very round and smooth in most cases.

We'll again start by adding more resolution, then defining the shape a little more, forming the end of the clavicles, and making a few smaller tweaks to the silhouette, such as making the outer contour a little flatter and slightly straightening the top:

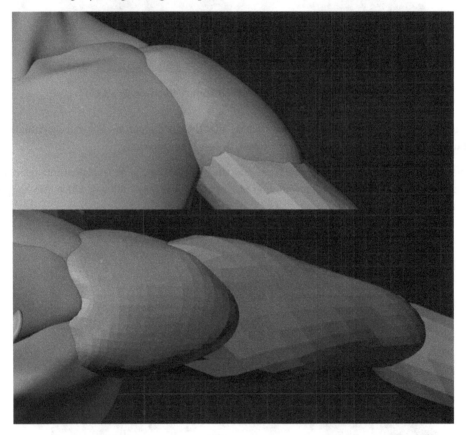

Figure 8.32 – Shoulder shape refined

If you want, you can make subtle separations between the different parts of the **deltoid** muscle using the **Clay Strips** and **Draw Sharp** brushes, although they're not visible on most people. We won't use separations on our models.

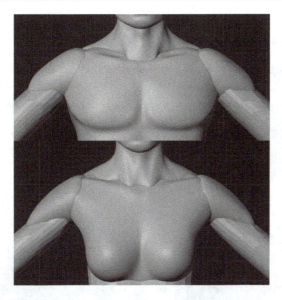

Figure 8.33 – Male and female shoulders sculpted

Notice how the female shoulders are slightly rounder relative to the male model's.

Moving on to the biceps and triceps, we can start by adding volume to the muscles using the **Draw**, **Inflate**, or **Blob** and **Clay Strips** brushes, tweaking the contour using the **Grab Brush**, and using the **Draw Sharp** brush to define the muscles a bit more. A reference is also extremely important here.

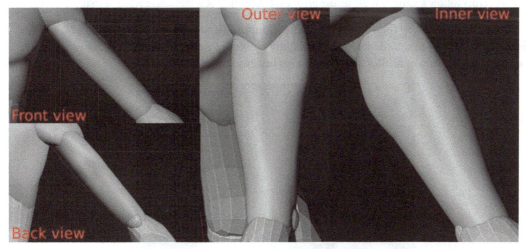

Figure 8.34 – Male biceps and triceps

Notice how in both outer and inner views, there's a subtle division between the biceps and triceps, and since in this case, they are relaxed, it tends to be less defined, though the extended arms make it so that the triceps have slightly more volume.

You can tweak the shape and definition of the muscles according to your taste depending on how strong you want the model to look.

For the female base mesh, the muscles remain the same, but they usually are slightly smaller and less pronounced.

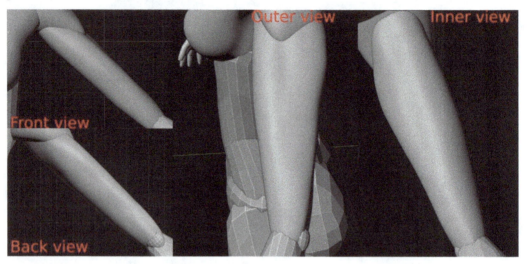

Figure 8.35 – Female biceps and triceps

The forearms follow a similar logic as the arm, except rotated at a 90° angle relative to it.

We'll start by building up the volume for the **brachioradialis** muscle, which is the biggest, and the **flexors**, sitting right next to it and having a slightly visible separation from the **brachioradialis**. For this, we'll use the **Inflate** or **Blob** and **Clay Strips** brushes.

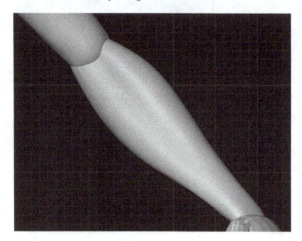

Figure 8.36 – Brachioradialis and flexors shaped

If you want, you can sharpen both the muscles and/or the division between them using the **Draw Sharp** brush.

Now for the **extensors**, which are more visible at the sides and back, we can use the same brushes but with a smaller size and strength, since those muscles are smaller.

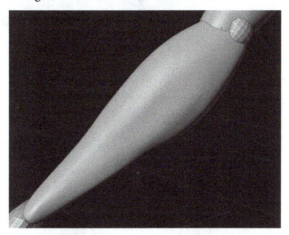

Figure 8.37 – Extensors shaped

The forearm muscles are less visible, even for reasonably athletic people, and especially when relaxed. Here is a picture showcasing two angles of the forearm:

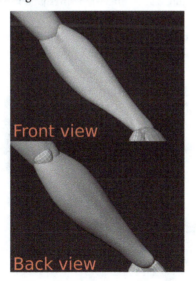

Figure 8.38 – Forearms shaped

For females, the muscles are usually smaller and even less prominent.

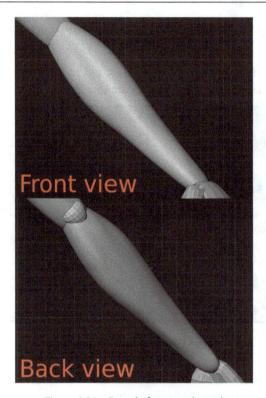

Figure 8.39 – Female forearm shaped

We can also shape the elbow, this time without adding resolution because it's such a small part. What we will do though is make it cover a bit more of the back since now we have mostly finished the arms. We'll turn it into a trapezoid shape, extending its sides outward. This goes for both of our base meshes.

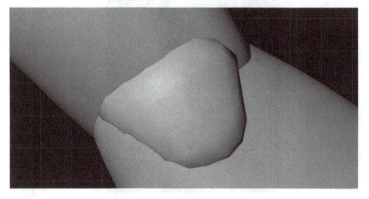

Figure 8.40 – The elbow shaped

Now let's leave the arms for now and work on the abdomen.

The abdomen

The abdomen is another part that has a great amount of variety from person to person, especially when comparing men and women, mainly due to overall muscle mass, and therefore must be handled with extra attention to references. We'll start with the male abdomen.

Since we're aiming for a more athletic look, the muscles in this area will be more defined.

As always, we'll start by increasing the resolution, and then we'll use the **Clay Strips** and **Draw Sharp** brushes to define the abs more.

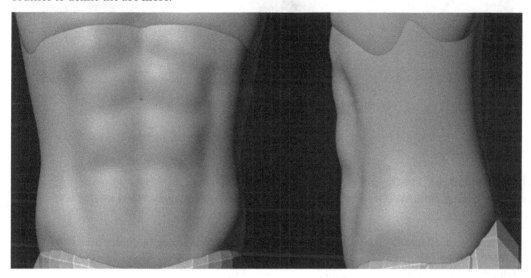

Figure 8.41 – Male abs

As you can see, we went for a squarer shape for the abs and defined the edges as well. Notice how the abs become slightly less wide as they extend down. You can add a belly button if you want, but it's not necessary for now.

Now we'll proceed to add further detailing around the abs, such as sharpening the muscles and tweaking the contouring a little. Here's a comparison between before and after these changes:

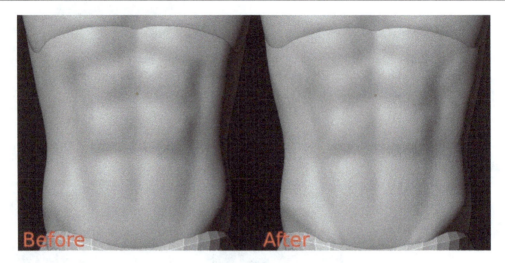

Figure 8.42 – Male abs further detailed

As you can see, the division between the muscles was slightly sharpened and the lower part was made a little thinner.

The male abdomen is mostly done, so we can move on to the female abs before we make the middle and lower back of both. We'll show a different method to add the muscles of the area that could also work in the male abs.

Even athletic women usually have abdomens with less defined muscles, so our base mesh will reflect that. We'll start by adding some volume to define the area using the **Clay Strips** and **Draw** brushes and the **Grab** brush to shape the contour:

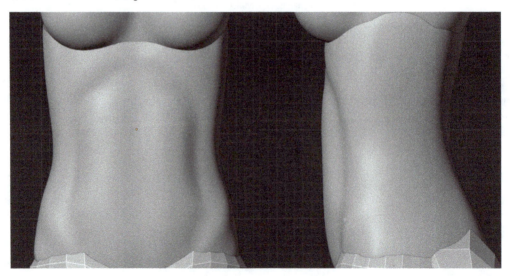

Figure 8.43 – Volume added to the area of the female abs

Now, we can use the **Draw Sharp** brush to add more definition to the abs, keeping in mind the lower muscle mass at play here.

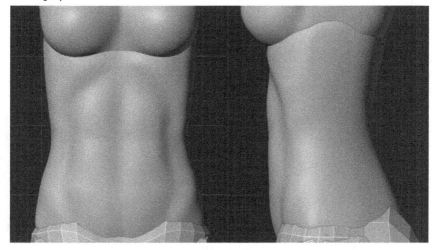

Figure 8.44 – Female abs more defined

Now, we'll proceed to make the middle and lower back of both base meshes.

We can start by adding the lower part of the **ascending trapezius** using the **Clay Strips** brush, extending to the middle of the back:

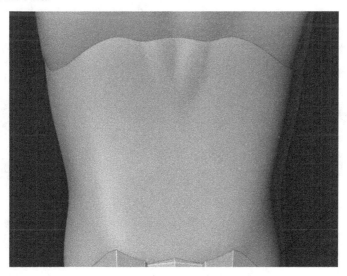

Figure 8.45 – The lower part of the ascending trapezius added

Keep in mind that it will lose its definition slightly when we add the rest of the details.

Now, we'll finish the lats since they extend all the way down to the lower back. Using the **Clay Strips** brush again, let's add some volume to this area. For a better visualization of this step, we'll add a before-and-after:

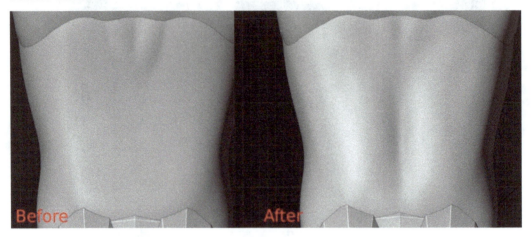

Figure 8.46 – The lower part of the Latissimus Dorsi added

As you can see, this muscle now protrudes more, and the back is less flat. A reference is useful for figuring out how this muscle looks at different levels of training, as we're not making someone extremely muscular.

For the female back, we can make everything we made for the male, making it thinner and more subtle. We'll also include a before-and-after for this:

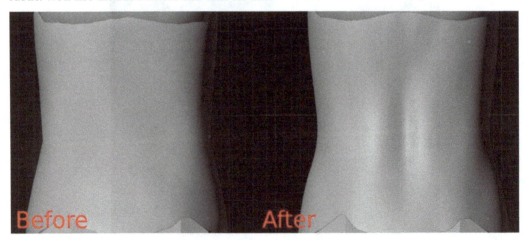

Figure 8.47 – The female back

Let's move on for now to the pelvic area.

The pelvic area

This part is less detailed compared to what we have done before now, so it tends to be simpler. Plus, it barely has any defining details, so that's nice.

We'll start with the male base mesh by adding more resolution and shaping it a little better using the **Grab** brush:

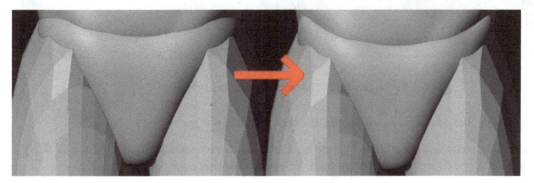

Figure 8.48 – Male pelvic area shape refined

The edges were raised a bit to fill the space left by the abdomen, and it will be smoothed after we join all the parts together.

Now, for extremely obvious reasons, we'll have to add volume to the lower part of it, which can be done using the **Blob** or **Inflate** brushes. It doesn't need to be detailed though, as we're not working with genitalia.

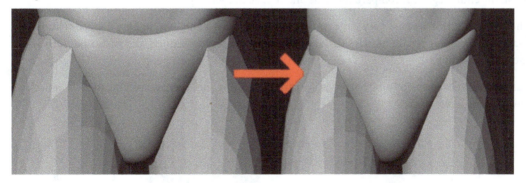

Figure 8.49 – Volume added to the male pubic area

You can add as much volume as you want; nobody will judge you.

For the female pelvic area, we also won't do much, just some tweaking of the shape to fill some potential space left by the abdomen:

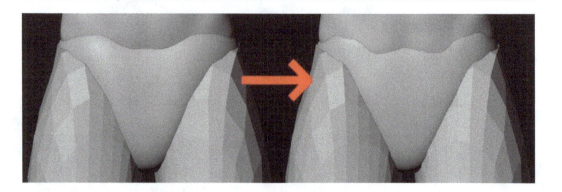

Figure 7.50 – The female pelvic area shaped

In this case, the sides were raised slightly, and the lower area was pushed back a bit.

Now, the gluteus has little work to be done since it usually is very round and smooth, so we'll skip it for now. Maybe flatten the sides a little bit if you want, but it's not essential at the moment.

We'll move on to the legs, which are more important and can give us better insight into how the gluteus should look.

The legs

It's just as tricky to make the legs look decent as it is to do so for the arms, but usually, the muscles are a bit more defined, especially in athletic people. Let's start with the male model's leg, and once we finish the entire leg, we'll compare the male and female model's legs.

The muscles in this area are very large, as discussed previously, and tend to have a more cylindrical shape. We have three main muscles in the **quadriceps**: the **rectus femoris**, the **vastus lateralis**, and the **vastus medialis**.

Using the **Clay Strips** and **Inflate** brushes, we can start adding some general volume to those muscles, always using a reference.

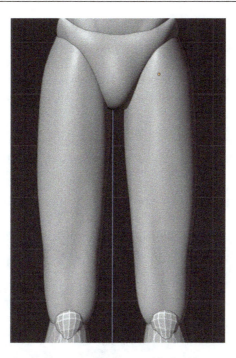

Figure 8.51 – The general volume added to the upper leg

Notice how the muscles now have a slight distinction from one another. We can now refine the shape of these muscles and sharpen their contour using the **Draw Sharp** or **Crease** brushes. Remember that the **vastus lateralis** (outer muscle) extends to the sides of the thigh, hence the name.

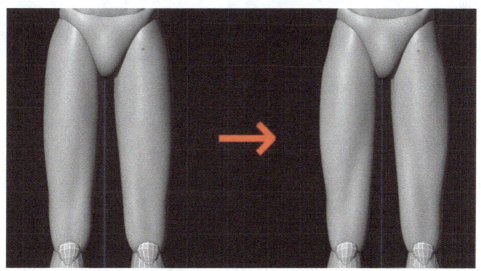

Figure 8.52 – Muscles further shaped

Notice how the muscles in the middle/upper portion are puffed up slightly, and the highest point of the leg has been pushed inward slightly, where the muscle has not yet come to its full volume and there are no bones. The creases between the muscles have also been sharpened a bit. Also notice how the **rectus femoris** (middle muscle) and the **vastus lateralis** muscles have very little distinction in our model, almost blending visually. That's the case for most people, although it can be very apparent depending on how strong you are.

Now, let's make the back of the upper leg. We have three main muscles to worry about: the **biceps** femoris, the **semitendinosus**, and semimembranosus muscles. They all extend from under the gluteus to the back of the knee.

Like with the front of the thigh, we'll start by adding more volume to the muscles using the **Clay Strips** brush:

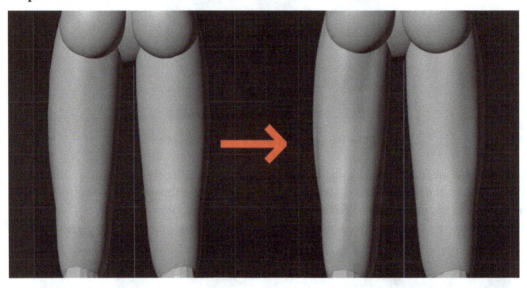

Figure 8.53 – Volume added to the muscles of the back of the thigh

Notice how the **semitendinosus** and the **semimembranosus** muscles also have little distinction from one another, making them look like one larger muscle.

Now we can refine the shape and definition of each muscle, tweaking the overall shape and sharpening the creases.

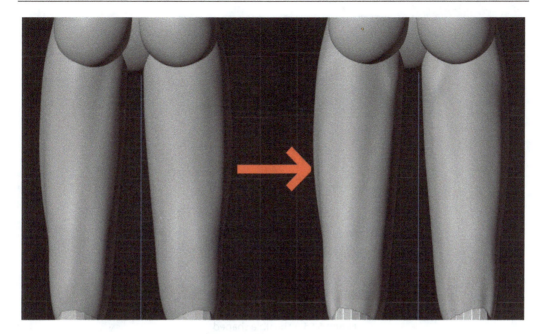

Figure 8.54 – The back of the thigh refined

Notice how the muscles gained slightly more volume, and how, near the knee area, the distinction between the muscles got sharper. That's because, near the knee, the muscles get thinner and are nearer to the skin than on the upper thigh, for example, where more fat accumulates. Remember that the visibility of the muscles in this area is generally lower than on the front of the thigh.

Let's move on to the lower leg for now.

The lower leg tends to be easier, especially since its main element is the calf, composed only of the **gastrocnemius** muscle and, since the front is mainly shaped by bone, it will have fewer details to worry about.

Let's start with the front since it's the easiest part. We'll shape the tibia (bigger lower leg bone) using the **Draw** brush since it's rounder:

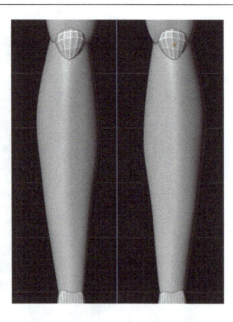

Figure 6.55 – The tibia shaped

Now, we'll increase the volume of the outer side of the lower leg slightly using the **Clay Strips** brush since the muscles responsible for some movements of the foot are located there.

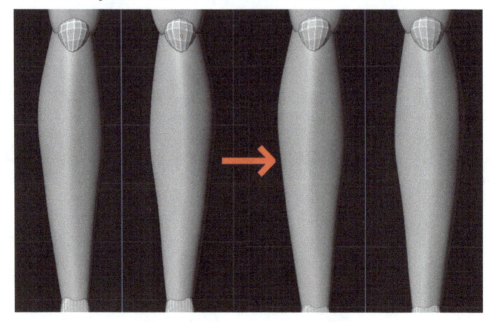

Figure 6.56 – Volume added to the outer portion of the lower leg

It's a really subtle addition, and in this case, can be perceived when comparing the diffusion of the highlights between them. Notice how the brighter part in the middle looks sharper on the left side and more diffused on the right; that's a result of adding volume to that area, therefore reducing the sharpness.

Now, we'll start shaping the calf. It is one muscle with two fibers, and there can be a significant separation between them depending on how strong the model is intended to look.

We'll start by adding more volume to the area and tweaking the shape using the **Clay Strips** and **Grab** brushes:

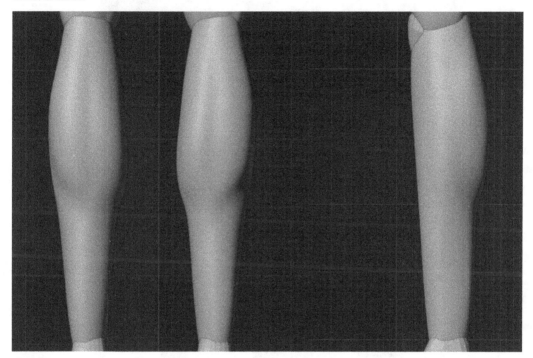

Figure 6.57 – Volume added to the calf

Notice how the muscle on the inner portion protrudes at a lower point than the muscle closer to the outer side and how the gastrocnemius muscle is not visible in the lower leg.

Now, for the lower portion, we can make the Achilles tendon by carving into the sides of the leg using the **Clay Strips** brush and shaping the tendon using the **Grab** and **Draw** brushes:

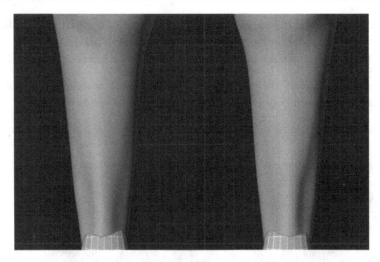

Figure 6.58 – The Achilles tendon added

Now, if you want, you can go back and tweak the shape of the muscles and sharpen or smooth the separation between the muscles according to your taste, always with a reference at hand. In our case, we decided to slightly sharpen the crease between the muscles and remove some of the muscle volume from directly under the **gastrocnemius**:

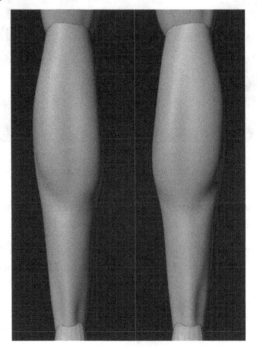

Figure 6.59 – Calves finished

Now, with the legs mostly finished, we can shape the knee. Using the **Clay Strips** brush, we can remove volume from both sides of the knee, leaving the middle portion more prominent, and then, using the **Grab** brush, we'll pull the lower part down to make the knee a little taller. We can also shape the kneecap using the **Grab** brush.

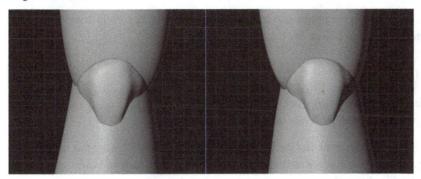

Figure 6.60 – Knees shaped

Now with the male model's legs finished, we can sculpt the female model's legs. These have the same muscles but they are less visible and the contour is usually curvier at some points. Let's have a look at the differences:

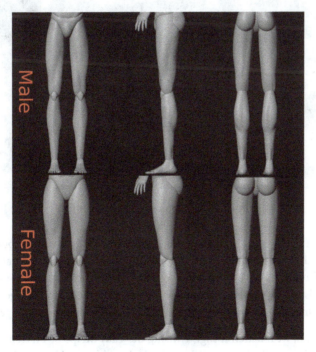

Figure 6.61 – Male versus female models' legs

Women's legs tend to have less muscle definition too, and the thighs are thicker and rounder most of the time, especially around the waist area due to the larger pelvis. The inner part of the thigh is curvier as well in our model, but sometimes it's possible that it is straight, depending on muscle training and fat accumulation. That's why we always keep a reference of the body type we're going for.

Now, we're only left with the hands and feet to refine before we join everything together and make each base mesh one single object. We'll start with the hands.

The hands

The hands between models also share a lot more similarities but are a complex part of the body, so we need to pay extra attention to our references.

We'll start by shaping the front and back sides of the palm first, increasing the resolution, and start adding some volume to the edges of the hand using the **Inflate** brush and carving the middle of the palm using the **Clay Strips** brush.

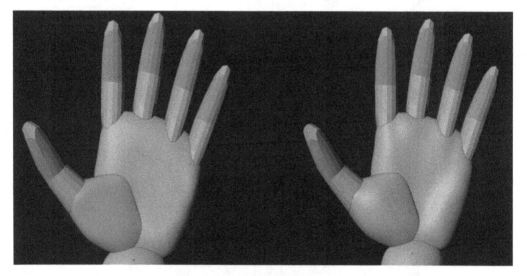

Figure 6.62 – Volume added to the hand

Notice how the middle of the palm is flatter while most of the edges have been puffed up a bit, and the base of the thumb has also been inflated. A good tip for this step is to look at your own hands and use them as a reference too. We won't add the little wrinkles we usually see in our hands since it would take hundreds of thousands of faces to properly sculpt them, and our goal with this is not a perfectly realistic human. We'll now move on to the back of the hand, which is generally simpler.

For the back of the hand, we'll flatten it slightly using the **Clay Strips** or **Flatten** brush, then using the **Draw** or **Blob** brushes, we'll add some volume to the knuckles, and lastly using the **Draw Sharp** brush, we'll add the tendons at the base of each finger:

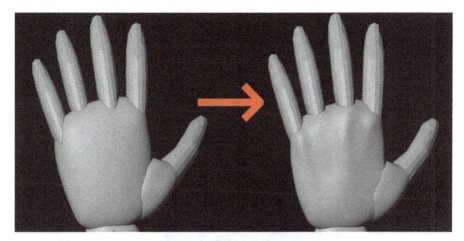

Figure 6.63 – The back of the hand refined

Notice how the tendons don't extend at a 90° angle, but rather converge into the wrist, becoming less and less visible as they get close. You can tweak that visibility according to your taste and/or goals.

Also see how the knuckles follow the arch of the fingers, and that not all of them have the same visibility. Depending on the level of realism, some of these details can be omitted. The visibility of these elements can also vary with age and weight. Again, use a reference for the specific body type you're going for.

Now, onto the fingers – we'll start by adding the joints using the **Draw Sharp** brush and tweaking the shape of the tip to make it thicker:

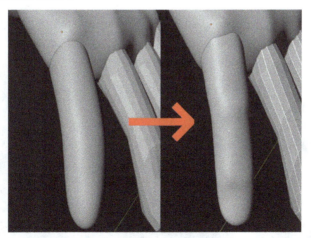

Figure 6.64 – A finger refined

Note that the joints appear on both sides of the finger except for the fact that while the back protrudes outward, the front protrudes inward. Both are subtle in most cases. If necessary, you can use the **Flatten** brush to slightly flatten the tip of the finger if it looks too round.

If we want, we can trace a fingernail, but it's unnecessary in our case.

Now, we'll duplicate that finger and then position and scale the duplicates on the correct spot for the other fingers, even the thumb:

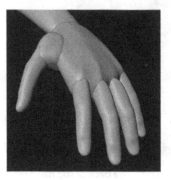

Figure 6.65 – The finished hand

Now, if you're creating a female character, keep in mind that the fingers tend to be thinner on women, and the joints and knuckles are slightly less apparent. Now, there's only one thing left.

The feet

Now, for the last body part: the feet. They are simpler than hands and have less fine-tuning to be done, so let's start.

Let's shape the back of the feet first, starting with the Achilles tendon, which extends to the heel. We'll use the **Draw** and **Clay Strips** brushes to shape it, carving the sides and accentuating the tendon itself:

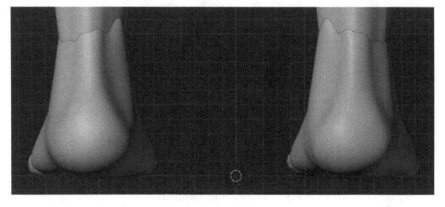

Figure 6.66 – The Achilles tendon shaped

Notice how the tendon appears to get wider as it approaches the bone located on the heel.

Now, we'll make the ends of the tibia and fibula bones, which are visible in most people as bony structures protruding out at the sides of our feet. For them, we'll use the **Blob** or **Inflate** brushes since they are rounded:

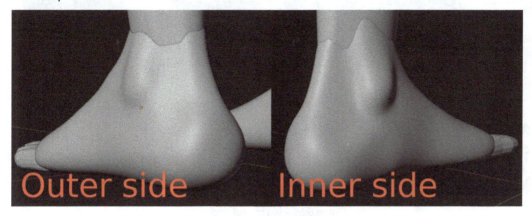

Figure 6.67 – The base of the tibia and fibula bones added to the feet

Usually, the protrusion on the inner side of the foot is located at a slightly higher point than the one on the outer side.

Looking good – now we'll proceed to make the top of the foot, by using **Clay Strips** in **Remove** mode or the **Flatten** brush to make the surface less round, then adding the tendons using the **Clay Strips** and **Draw Sharp** brushes:

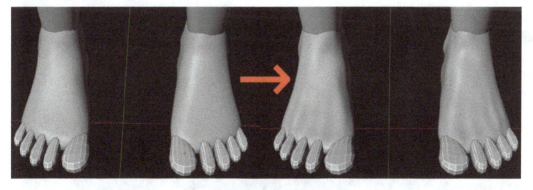

Figure 6.68 – The top of the foot shaped

Now onto the toes, which will have a similar approach to creating the fingers on the hands, except we'll also have to sculpt the big toe since it is significantly different from the other toes. We'll start with the big toe.

We won't add much detail to the toes since they're so small and not the focus in or case, so we'll just add the joint and flatten the tip a bit, both using the **Clay Strips** brush:

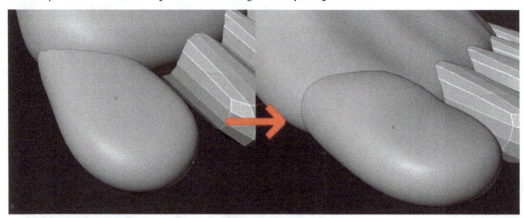

Figure 6.69 – Big toe refined

Notice how the base of the toe was pushed up a little to represent the joint. As always, feel free to add as much detail as you want/need such as adding nails or a different shape to the toes.

Now for the smaller toes, we'll shape just one, this time adding two joints instead of one:

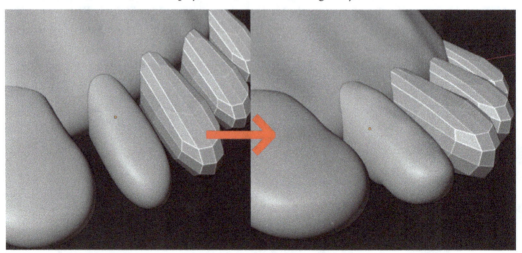

Figure 8.70 – Smaller toes shaped

Now, we'll finish the foot by duplicating the smaller toe and positioning it in the correct spots – here's the finished result:

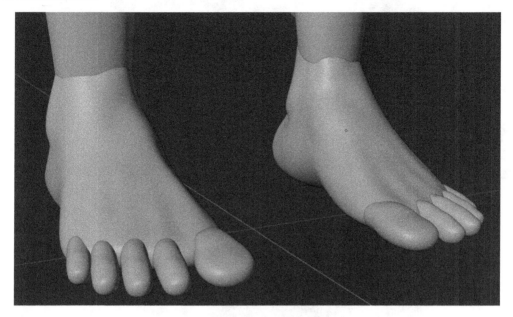

Figure 8.71 – Finished foot

Keep in mind that women generally tend to have smaller feet than men.

Congratulations! We're finished refining our base meshes, and now we can join all of the parts and then re-mesh them to make each base mesh one single object by smoothing the transitions between the formerly separate objects.

Before that, though, it's good practice to go back and make any adjustments to the shapes and proportions, taking a close look at each portion of the model to spot potential flaws that could have been looked over in the process, with reference in hand.

Let's have a look at how the base meshes look in a full-body picture of our base meshes:

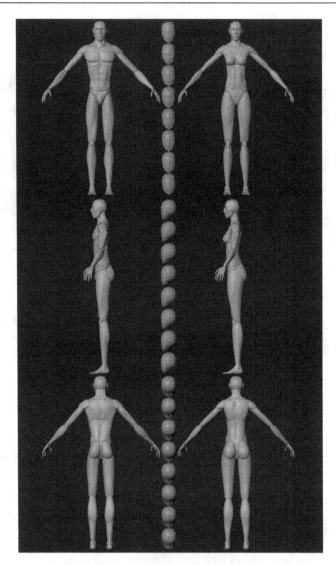

Figure 8.72 – Full-body refined base meshes

Merging the different parts

Once we're happy with the results, we can proceed to apply every modifier (the **Multiresolution** and/or **Mirror** modifier, for example) to every separate object since for them to be joined correctly, there needs to be no modifiers added to any of the objects composing the body; otherwise, they might conflict and make the mesh look weird. Before that, though, it's recommended to save the progress and make a copy of the body parts, in case anything goes terribly wrong.

Now in **Object Mode**, after applying all modifiers, we can select all the separate objects that make up each of the base meshes (except the eyeballs, since usually we don't want them joined with the body), and join them together using *Ctrl + J*. Then, we'll enter **Sculpt Mode** again and use **Remesh** on the joined objects using *Shift + R* to define the resolution and *Ctrl + R* to apply that to the mesh; we went for around 0.01 m for **Voxel Size**, for minimal loss of detail. Once you do that, you should see that the edges between the former separate objects have merged, and you can use a combination of brushes to smooth the edges that are still harsh. As an example of how to do it, let's have a look at the hands, which were made up of a lot of separate objects previously:

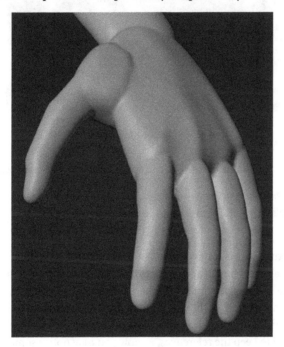

Figure 8.73 – Remeshed hand

Notice how the geometry of the fingers is now merged with the hand. Be careful, though, as some fingers, toes, and other parts very close to one another might be merged together. If that's the case, undo the re-mesh using *Ctrl + Z* and spread them out a little more.

With this, we can start smoothing the edges, which are still quite harsh. Using mainly the **Smooth** and **Clay Strips** brushes, we can start going over the edges, smoothing them out to make an organic transition between palm and finger. Remember to keep **Symmetry** always *enabled* from now on to avoid future headaches.

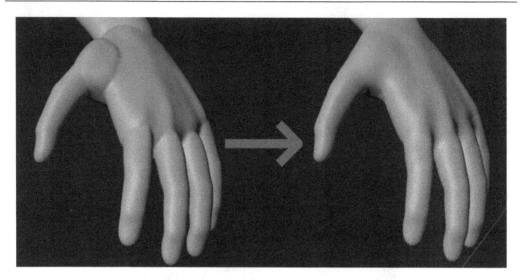

Figure 8.74 – Hard edges smoothed

See how this looks much more natural, without the edges left by the original remeshing.

Sometimes, though, the **Smooth** brush won't be able to create a smooth transition alone, usually leaving some sort of valley where the transitions happened between objects. When this happens, we can add a little bit of volume there using the **Clay Strips** brush, for example, then smoothing it again.

Perfect – now, we repeat the process for every part of the body that has hard edges remaining where there shouldn't be. As always, feel free to add some tweaks to the sculpture, as now we have the entire body in one mesh and are able to sculpt in areas where previously we couldn't, such as joints. There's not much more to add to this process, as the specific actions necessary for smooth and natural blending can vary greatly from model to model – it's really a matter of what gives out the best results. Even at this stage, though, it's good to keep the references, as now you may notice some pieces of the anatomy that look weird that you previously didn't, especially near the joints.

Keep in mind that with this smoothing, the body parts may become slightly thinner, so we might have to fix that afterward.

Once we're done, we should have these as our finished base meshes:

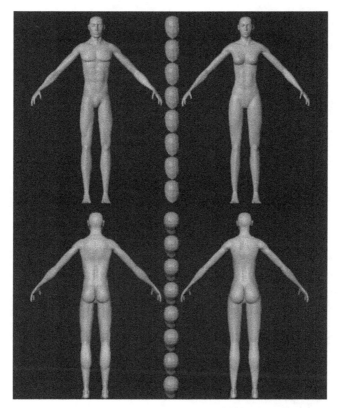

Figure 8.75 – Finished base meshes

Perfect – both our male and female base meshes are looking good. Now, if you want, you can use the **Grab** and **Inflate** brushes to pull both parts of the gluteus closer together, but it won't make a significant difference unless that's an area of focus and/or will be closer to the viewer.

Even though they look good, they still need to be heavily optimized, as each of them currently is at around 1 million polygons, which is unsuitable in most cases, especially with the geometry left by the remeshing operation we used to merge the parts together. But hey! The hard part is over!

Summary

In this chapter, we covered how to refine the block out made in the previous chapter, making sure that we have reasonably precise anatomy while sculpting in order to avoid our base meshes looking wrong. We also suggested which brush to use in each situation based on the shape of the body part and the type of detail we added.

In the next chapter, we'll cover how to deal with the current geometry, as it's far from perfect, as well as one way of unwrapping a humanoid character to prepare it for texturing.

9
Optimizing the Base Meshes

Now that we have successfully completed our base meshes, we need to optimize their geometry in order to be able to do things such as unwrap, add textures, and pose them (which will be covered in the next chapter). The thing is, unlike with models made with regular poly modeling, the geometry usually left by sculpting is too messy and dense to be dealt with by just dissolving edges or merging vertices. Rather, we'll recreate the model from scratch with much cleaner geometry, using the high-density mesh as a base to lay the new polygons on, in a process known as retopology.

In this chapter, we'll cover the following topics:

- The basics of good topology on a sculpt
- How to retopologize a humanoid character
- How to unwrap the new geometry

By the end of this chapter, you will be able to retopologize humanoid characters with clean geometry. Some of the principles covered extend to other types of characters too, such as animals.

What is good topology?

At the beginning of this book, we covered how to optimize a model, so some of the principles of good geometry are already known. But when talking about characters, more things have to be taken into consideration, such as edge flow and poles, which are both defining factors when dealing with this kind of model and will be covered soon.

To start, decent topology is composed mainly (if not only) of quads, since they give the best results when it comes to a mesh that is supposed to deform, such as joints and the face when doing facial expressions. We should ideally keep the whole mesh as quads, especially in areas of high deformation.

Now, with quads in mind, we can talk about edge flow.

Edge flow

Edge flow basically means the direction that a string of connected faces is pointing in. The only polygon that can have edge flow is a quad, and each quad has two directions for edges to "flow" when added, horizontally and vertically (edge flow will always be represented with a yellow line here):

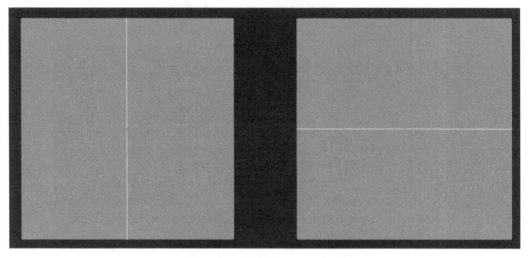

Figure 9.1 – Edge flow on a plane

Now, if we deform this quad on any axis without deleting any vertices, the edge flow will follow the shape:

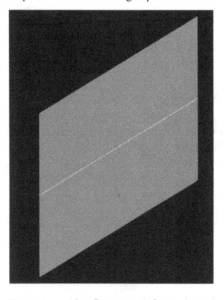

Figure 9.2 – Edge flow on a deformed quad

That means that however our mesh is laid down, there will be edge flow as long as it's composed of quads.

In a character, we need to keep this flow as clean as possible, following the shape of the mesh in order to properly keep the detail with as little geometry as possible. Let's have a look at very simple examples of good and bad edge flow around a mesh. We'll use a sphere to illustrate this:

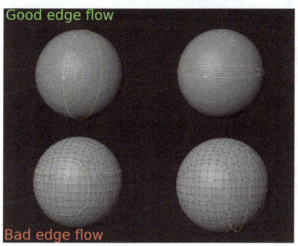

Figure 9.3 – Good and bad edge flow

Notice how a good edge flow is smooth and makes a perfect loop around the sphere, while in the bad example, it's all crooked, and one of them even spirals around into a different flow when the loop is placed in one of the directions, which is typically not ideal. Unfortunately, though, the geometry left by sculpting often tends to resemble the bad example, but way, way worse.

Another thing to keep in mind when retopologizing a mesh is poles.

Poles

Poles are any vertex that has more than four edges connected to it. This is something we'll all have to think about while retopolozing, especially with characters and meshes that deform:

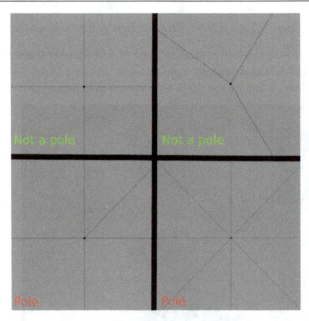

Figure 9.4 – Poles

Ideally, we'd only have four connecting edges in one vertex, but for the most part, it's impossible to completely avoid poles. However, we can try to control their placement.

As a rule of thumb, we should avoid putting poles in areas of high deformation, especially on the face (if it's going to be shaped or animated later on, of course), since they tend to cause pinching when deformed or on curved surfaces:

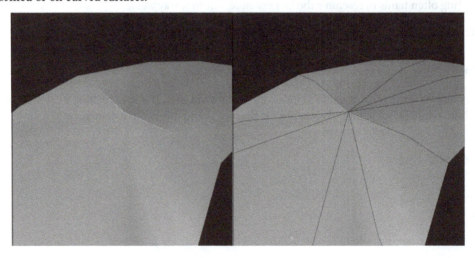

Figure 9.5 – Pinching on a curved surface caused by a pole

Notice how the surface appears overly distorted, almost like there's something pinching the mesh toward the center, where the pole is located.

Now, with this in mind, we can start the actual retopology process of our base meshes.

Setting up the retopology mesh

The actual setup of the retopology is relatively easy, since we'll only need a mesh that sticks to the surface of the high-density sculpt.

To start, we'll add a simple plane and scale and position it near the area we want our retopology to start. In our case, this is the eyes:

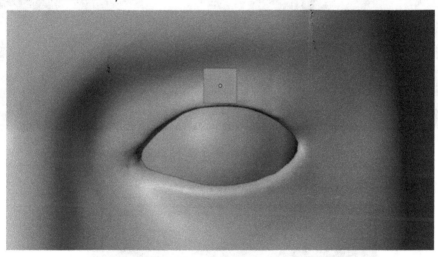

Figure 9.6 – Quad placed on top of the upper eyelid

Make sure to check that the normals are facing the right way.

Now, we'll proceed to make it stick to the surface of the mesh we want to retopologize, by using a **Shrinkwrap** modifier, which is located in the **Deform** column and will make our new mesh stick to the surface:

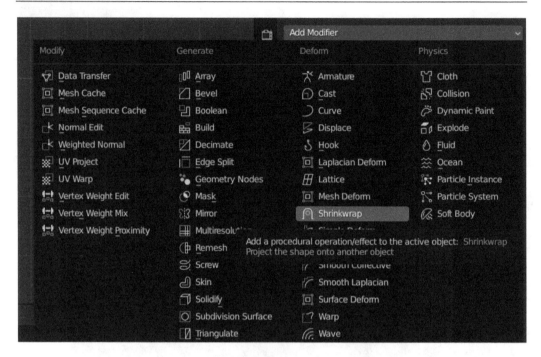

Figure 9.7 – Adding the Shrinkwrap modifier

Upon selecting the modifier, you'll be presented with this:

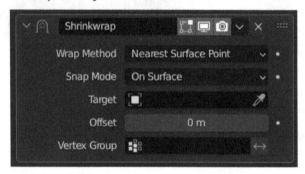

Figure 9.8 – Shrinkwrap modifier

Most of the settings are self-explanatory, and the default settings work the best in our case. The only thing necessary for us to do is to select the target, which will be the base mesh. To select it as a target, we can click the eyedrop icon to the right of the **Target** option, then click on the mesh we'll retopologize. Instantly, we'll see that the quad we added is snapped onto the surface of the eyelid:

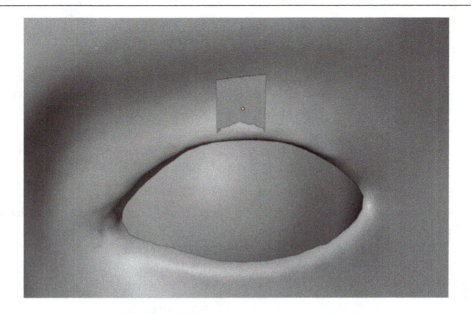

Figure 9.9 – Quad snapped onto the mesh

Upon selecting the base mesh as the target object in the **Shrinkwrap** modifier, a new option will appear in it, the **On Cage** setting:

Figure 9.10 – On Cage setting

We'll activate it, since this makes it so that the actual vertices appear to us as snapped into the mesh, which wouldn't happen otherwise since modifiers work by calculating/displaying the changes while storing the original mesh. Once in a while, though, it is good practice to duplicate the modifier with *Shift + D*, then apply the first one, to commit to the changes done previously. This can help avoid some weird behavior from the modifier while retopologizing.

Since our base mesh is perfectly symmetric, we'll also add a **Mirror** modifier with **Clipping** enabled to the plane we added for retopology. That way, we can only work on half of the mesh.

At this point, you might have noticed that it has become hard to see what we're doing when the geometry is directly on top of the high-poly surface. For this reason, we'll make the retopology object a distinct color and lower the opacity, while making sure it always appears in front of the high-poly mesh.

To do that, we'll do the following:

1. Change the solid viewport shading from **Material** to **Object** (*Figure 9.11*).

2. Check the **In Front** option under the **Viewport Display** menu.

3. Change the color of the object under the **Viewport Display** menu. We'll use a blueish tone (*Figure 9.12*).

Figure 9.11 – Changing the viewport shading option

Figure 9.12 – Changing the color of the object and making it visible through other meshes

Don't forget to lower the **Alpha** option while changing the colors to make the object slightly transparent so that we can also see the original high-poly mesh. We used a value of 0.5 for 50% opacity.

Now, if we look back at our quad, we'll see this:

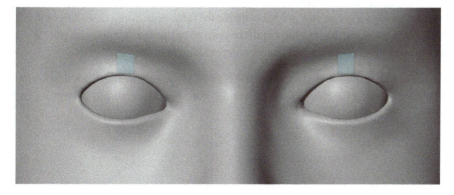

Figure 9.13 – Quad with better visibility

Perfect, now we can much better see what we're doing. We can now start retopologizing our base mesh.

Retopologizing the base mesh

With the setup successfully completed, we can actually start constructing the mesh. Now, in order to get a more efficient and better result, we won't start retopologizing big or detailed areas at once. Rather, we'll start by sketching in the main face loops to define the general edge flow of the mesh. We'll start with the head and work our way down.

Head

We'll start the head by extruding our face loop from the quad we initially added by pressing *Ctrl* and right-clicking where we want the extruded edge to be, with one loop around each of the eyes:

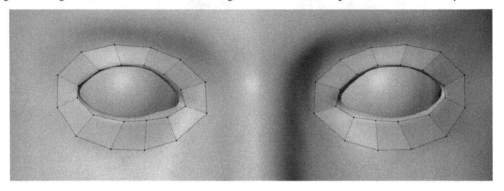

Figure 9.14 – Face loop added around the eyes

See how we only made the faces as big as the eyelids, in order to cover them by following their shape and forming a closed loop. You can merge the vertices using the *M* shortcut and fill the space between two edges using *F*, as covered in a previous chapter.

We'll keep the retopology mesh as low poly as possible for now, since doing this allows more control over the mesh when tweaking. It's also important to keep all of the faces at a similar size. We initially used 12 faces for this loop, though it will likely need tweaking later on.

Now, we'll proceed to add another loop around the eyes, but this time we'll make a single loop going around both eyes, following the shape of the face and eyebrows and meeting at the nose:

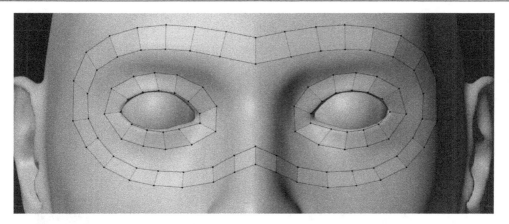

Figure 9.15 – Loop added around both eyes

The next loop will be just under this one, going around the nose and closing just under the mouth.

Make sure to somewhat match the placement of each face so it becomes easier to connect later when we start filling the gaps between the main loops we're making now:

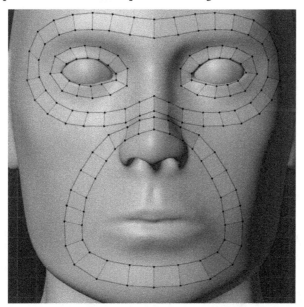

Figure 9.16 – Loop added around the tip of the nose and mouth

The next important loop goes around the lips, again following their shape:

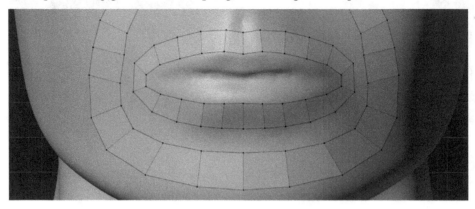

Figure 9.17 – Loop added around the lips

Remember that at any point we can add or remove edges, as well as tweak their placement on the face, as this initial number of faces may not hold up when connecting the edges.

From now on, the loops that are not too obvious will be highlighted in yellow.

Alright, the next loop is connected to the mouth loop. Starting near the edges of the mouth, this loop connects into the loop around the nose and mouth, goes up over/around the cheekbone and eyes, and ends at around the middle of the forehead:

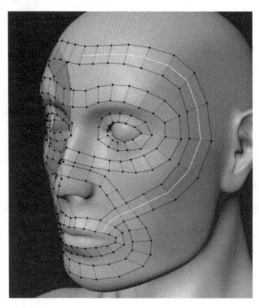

Figure 9.18 – Loop added around/over the cheekbones

Again, make sure to try and match the number of faces of the nearby face loops and remember that areas with less detail can have bigger faces.

Now, the next important loop starts at the chin, goes around and contours the jawline, and extends until the top of the head:

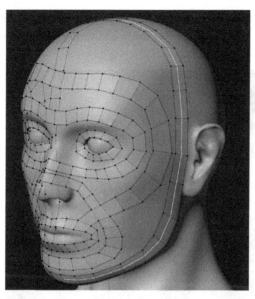

Figure 9.19 – Loop around the jawline added

Looking good. Now, the last face loop we'll add is around the ears:

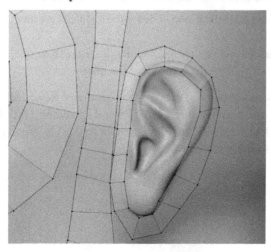

Figure 9.20 – Loop added around an ear

This loop was added around the part of the ear that is connected to the head.

Good. Now, with all of the most important face loops added, we can start tweaking the placement of each edge, in order to better match the number of faces in the nearby loops. We can also do it while connecting the loops, which could be a better method for some people.

Perfect. Now we can start actually connecting the loops to each other, starting from the middle of the face with a string that crosses the entire face vertically, connecting all of the loops in the area starting with the loop on the top of the head:

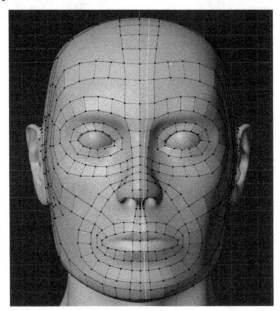

Figure 9.21 – Face loop added in the middle of the face

Notice how the faces get narrower when they reach under the nose. That's fine, since those faces follow the detail in the area, which is thinner than the rest of the nose.

Now, we'll fill in the area around the eyes, tweaking the geometry we added earlier if necessary, in order to add the new faces in a clean manner. Even if we end up adding a little bit of extra geometry, cleaning and smoothing it afterward will generally be easier, so don't be afraid to add more geometry to the loops created. It's almost like a puzzle.

From now on, the new connections added will be highlighted in red, as it may not be obvious where they're placed:

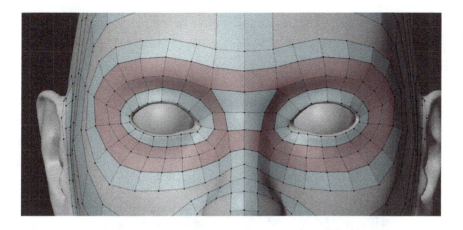

Figure 9.22 – Eye loops connected

As you can see, two new loops were added: one around each eye and one following the shape of the outer loop. For this to be possible, the geometry from the loops around the eyes might need some tweaking, but this can vary from model to model. Notice as well that a pole was created near the middle of the nose bridge.

Keep in mind that if you do need to add more geometry, you'll need to tweak the edges so as to evenly distribute the new geometry around. You can use either proportional editing or the **Smooth** brush in **Sculpt Mode** for that. You can also use a topology reference if you get stuck somewhere, as there are loads of different examples online.

Now, we can fill the rest of the space between the face loop around both eyes and the loop on the cheekbone. Don't forget to tweak the geometry in that area to make the loops smoother:

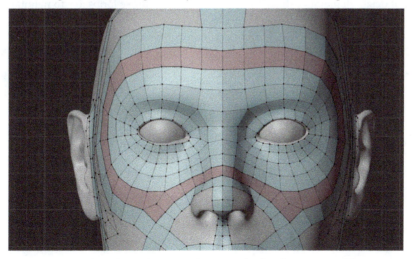

Figure 9.23 – Complete connection between the face loops around the eyes

In this process, another five-edge pole was created near the cheek area. Most of the time, five-edge poles don't cause many problems when it comes to deformation, so don't get too caught up in those.

Next up, we can fill the gaps between the loops around the jawline and the mouth, still trying to keep the quad distribution as even as possible.

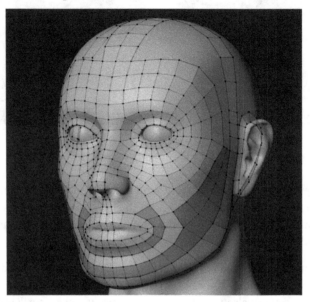

Figure 9.24 – Gap between mouth and jawline filled

Looking good. Next, we'll fill in the top of the head, which should be simpler to do since it's a surface with not a lot of detail. Remember that if you need to add or remove loops, try to maintain an even distribution with polygons that are roughly the same size.

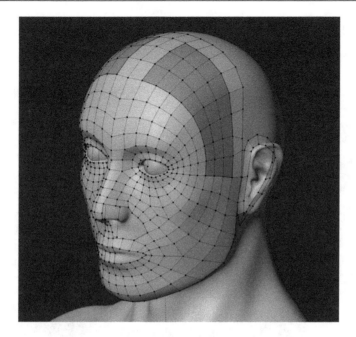

Figure 9.25 – Gap on top of the head filled

Now, with the bigger and flatter areas covered, we can go back to the mouth and nose, which can be trickier to deal with. Let's start with the mouth since it's a bit easier.

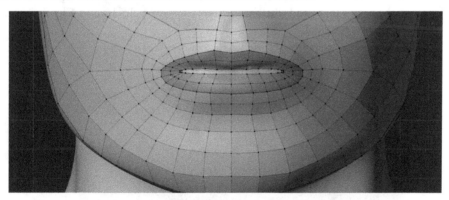

Figure 9.26 – Topology added on the mouth

Notice how a gap was left where the lips meet. Unconnected vertices should be left for facial expressions when the mouth needs to be opened. If you don't want or need the mouth to open and close, you can merge the vertices at the center and a triangle will most likely form at the edges of the lips, or you can add a loop on the edges of the lips and add two rows of faces closing the gaps, keeping the geometry in all quads. We don't cover facial expressions or animation in this book, so we'll close the gap.

Now for the trickiest part: the nose.

We'll start by adding a face loop directly around the nostrils, at the connecting point between them and the face:

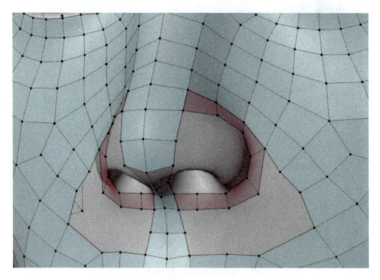

Figure 9.27 – Nostrils contoured

Now we can start filling up the hole between the nostrils and the top of the mouth, keeping in mind the number of faces on the nearby loops and whether there are excessive poles around. Closely inspecting the area we're working on is essential.

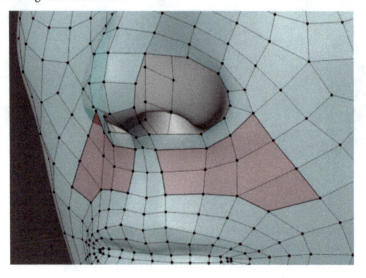

Figure 9.28 – Gap on the upper part of the mouth filled

Now, the only thing left is the actual nostrils. We'll start by adding a face loop around the bottom edges of each:

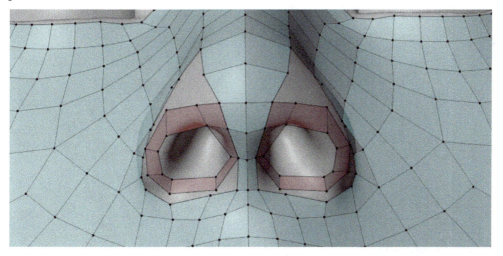

Figure 9.29 – Face loop added around each nostril

Usually, when there is a round shape in the sculpt, there will be a loop around it.

Now, we can fill the space left on the sides of the nose, still going around the nostrils and meeting at the center:

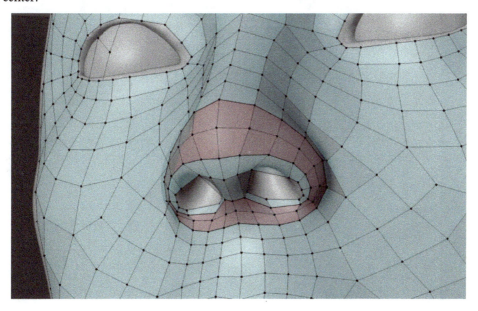

Figure 9.30 – Edges of the nostrils fully covered

Now, for the inside of the nostrils, we can select the loop closer to the inside, and instead of hitting *F* to fill it with one face, we'll hit *Ctrl + F* to open the **Face** menu, then select **Grid Fill** (this can also be done in the eye sockets if you need it):

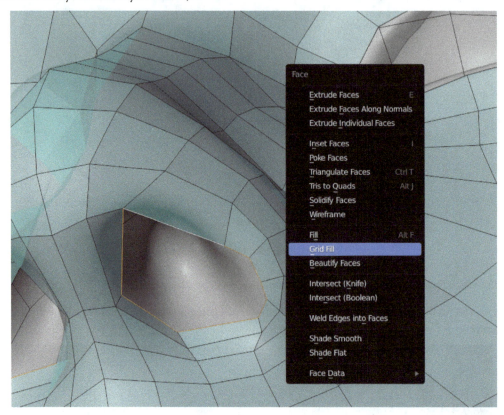

Figure 9.31 – Grid Fill

Now, when you select **Grid Fill**, the edge loop selected will be filled with faces, in a grid that can look different from model to model, depending on the shape and how many vertices the selected loop has.

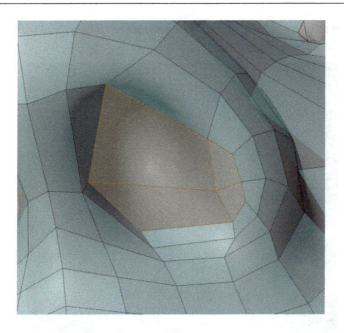

Figure 9.32 – Grid fill generated

Nice, it created an all-quad fill, but it doesn't go very deep into the nostril. For that reason, we can inset the faces generated one or two times, so the **Shrinkwrap** modifier can reach deeper:

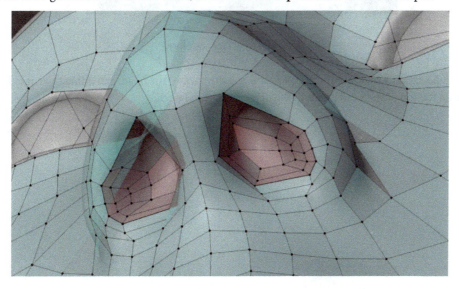

Figure 9.33 – Inside of the nostrils filled

With this done, all of the main parts of the face have been successfully retopologized. Here is the full new geometry of what we have done so far:

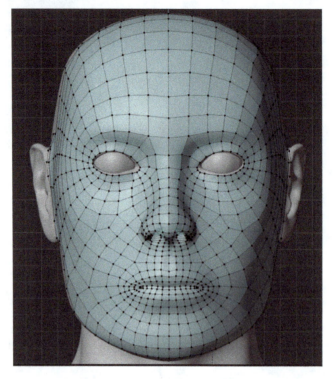

Figure 9.34 – New face topology

Looking good so far. Now we can start going around the back of the head, extruding the edges from the loop on the top of the head down to the upper part of the neck:

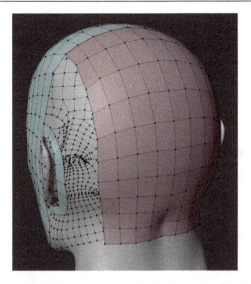

Figure 9.35 – New geometry added to the back of the head

Since the surface is not very detailed, the faces can be bigger, covering a larger area with less geometry. Also, since it doesn't have many deformities and different shapes like the face, it's way easier to add good geometry to it. Remember to still keep the faces relatively evenly distributed.

Now, we'll connect the loops on the back of the ear to the faces we just added, leaving a hole in the sides of the head:

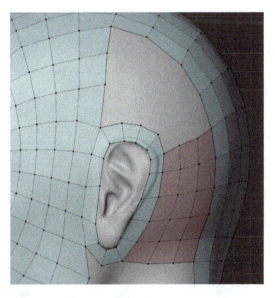

Figure 9.36 – Ear loop connected to the back of the head

Notice how the number of faces on each side of the hole left is the same as their counterpart; that is, the loop on top of the ear has the same number of faces as the top of the head, and the number of faces we added to the side of the head is the same as those we have on the connection between the ear and the back of the head. This sort of matching makes it extremely easy to fill out the space, as we can also use **Grid Fill** to fill such spaces uniformly in a grid. This technique can be used in quite a few situations where the surface is relatively flat and/or doesn't have a lot of detail that needs to be maintained when retopologizing:

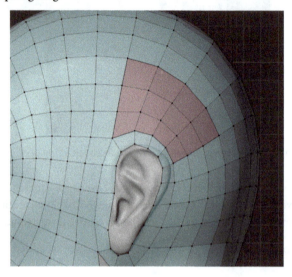

Figure 9.37 – Grid Fill used on the gap on the sides of the head

Now, we'll connect the faces on the back of the jawline with the ones on the upper neck. Make sure to also connect the faces of the bottom of the ear with those faces we add:

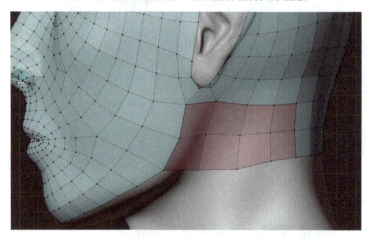

Figure 9.38 – Jawline loop connected to the faces on the upper neck

Now, the last thing left on the head is the ears. We'll start by extruding the back of the ear loop inward, in another ring covering the back of the ear:

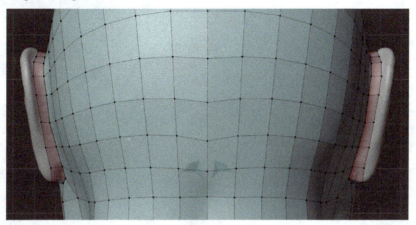

Figure 9.39 – Loop added to the back of the ear

We'll now extrude this loop one more time, extending the new faces to the sharper edges of the ear, and one last time, going toward the inner part of the ear until the next sharp edge:

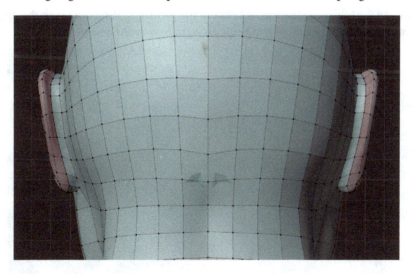

Figure 9.40 – Two more loops added to the edges of the ear

The inner ear can be a bit tricky to get right, so we'll pay a little more attention to it.

To start, let's cover the bigger cartilages inside the ear:

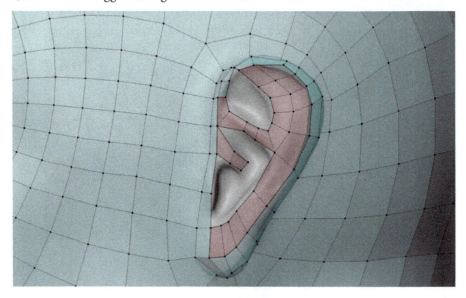

Figure 9.41 – Ear cartilages covered

Good, now we can start filling in the spaces between the cartilages. We'll start with the upper one, since it's the easiest, with only a simple grid being enough to cover the area nicely:

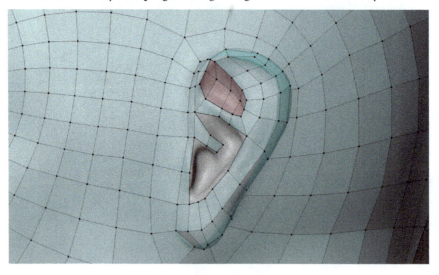

Figure 9.42 – Gap on top filled with a grid

Now, the bottom gap can be a little trickier, since it has a weird shape. First of all, we'll need more resolution, since what we have now is too little. But how do we add more resolution? Well, we can add a quad in a shape that resembles a diamond to the area, effectively doubling our resolution:

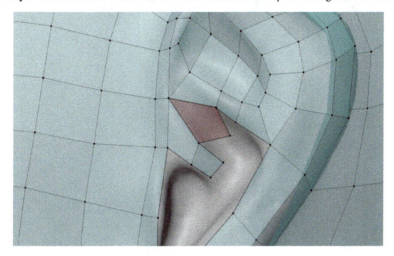

Figure 9.43 – Diamond-shaped quad added to increase resolution

Again, don't be afraid to add more edge loops if you need them, and remember to keep the new geometry as uniform as possible!

Good. Now that we have two edges to work with, we can connect the top one to the nearby edge face and start to make our way down the ear canal with the bottom one:

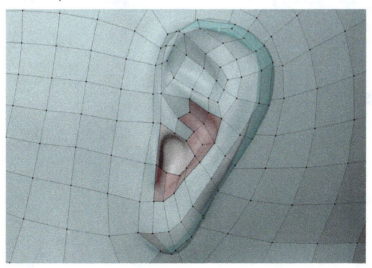

Figure 9.44 – Gap in the lower cartilage covered

Now, the only thing left is the actual ear canal, which we'll fill in a slightly similar way to what we did with the top cartilage:

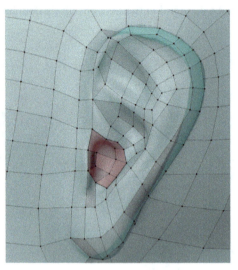

Figure 9.45 – Ear canal filled

Perfect, now we have a fully retopologized head with way fewer faces than before and a way cleaner geometry. Since the male and female heads follow the same logic, they can have a very similar, if not identical, retopology. Here are both fully retopologized heads:

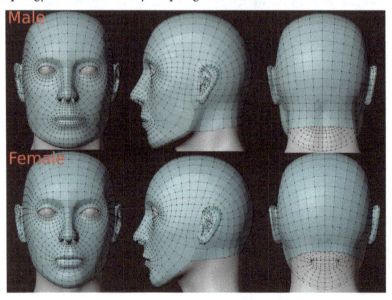

Figure 9.46 – Male and female fully retopologized heads

Now, you may notice that they have slightly different geometry, and that's completely fine. There's more than one solution that can work nicely. Even though the final result is not completely the same, both follow the same logic and have all of the same essential loops we covered.

You may also have noticed that the upper neck geometry on the female is slightly curved. If that's the case, it's not a big deal since we can smooth all of that geometry later.

Now, before we move on to the neck, chest, and back, it's worth mentioning that we don't need anywhere near as much resolution on those areas as we needed on the face, since they don't have much detail in our case. So, because of that, let's have a look at a few ways to decrease the resolution while still keeping an all-quad geometry.

How to decrease the mesh resolution

Decreasing the resolution for areas that need less detail is crucial to be able to have an optimized result. In order to reduce the resolution while still ensuring a decent all-quad topology, we can follow a simple logic that can be applied to many situations. Let's take a closer look.

Two faces to one

This method reduces the resolution by half and consists of connecting one face to another "diagonally", so that we get rid of an extra face. Let's demonstrate.

Suppose you have a string of faces that is two faces thick and you wanted to turn that into one face due to the level of detail of a nearby surface:

Figure 9.47 – Two-faces-thick loops

Now, to decrease this resolution by half, we'll first extrude both of the faces:

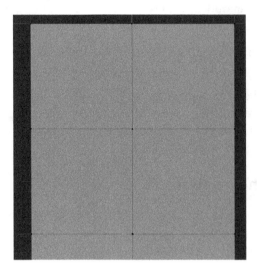

Figure 9.48 – Faces extruded

Then, we'll select the middle vertex and the upper outer edge vertex, then connect them with an edge using the **Join** tool (this can also be done by using the shortcut *J*), making a diagonal edge that crosses one of the new extruded faces, which is highlighted in white:

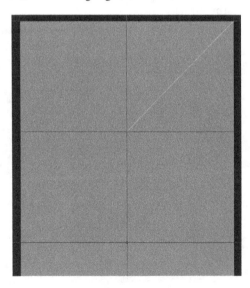

Figure 9.49 – Middle and upper outer edges connected

Now, the last thing to do is to dissolve the vertical and horizontal edges that make up that quad and you've got yourself an edge with half of the resolution:

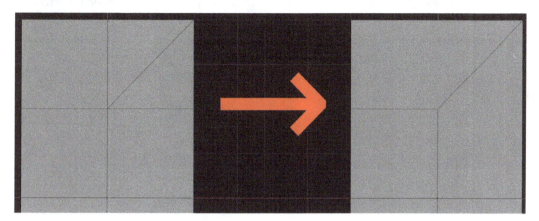

Figure 9.50 – Two-to-one face conversion

While keeping good topology, this also turns one of the face/edge loops 90°, which is not harmful most of the time, especially if we're working on a symmetric model with a mirror modifier, since the new loop will encounter its counterpart at the other side and turn another 90° (edge flow highlighted in yellow in the following figure):

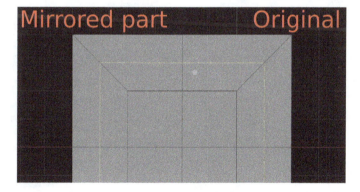

Figure 9.51 – Mirrored loop with two-to-one conversion

This figure also demonstrates a four-to-two-face conversion, as you can clearly see.

OK, now let's have a look at some other resolutions to be decreased. They will be covered quicker since the logic is essentially the same as this one.

Three faces to one

This decreases the resolution by three times and is very similar to the previous one, the only difference being that we connect the two vertices of the quad in the center to the outer edges of their respective sides:

Figure 9.52 – Three-to-one-face conversion

Unlike the two-to-one conversion, this one turns the loop a full 180°:

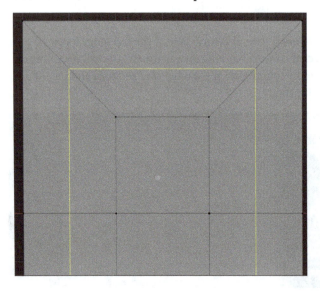

Figure 9.53 – Loop turned 180°

Five faces to three

This reduces the resolution by around 1.5 times and follows a similar logic than a two-to-one conversion, except instead of connecting the middle to the edges, we'll leave the middle face as is and connect the faces directly next to it to the edges:

Figure 9.54 – Five-to-three-face conversion

Additionally, you can space the two vertices left in the middle evenly to get more uniform faces when extruding from those edges:

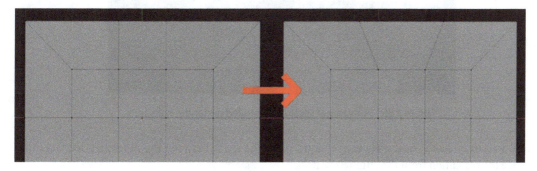

Figure 9.55 – Evenly spaced vertices in a five-to-three-face conversion

It's worth noting that in any of the conversions mentioned, the opposite can be carried out too in order to increase the resolution. Also, you can combine two of those methods to get different resolutions. So, we have a lot of control over the resolution we want for each part of the mesh.

Now, let's go back to our base meshes.

Neck

There's a reason for not covering the neck along with the head, even though it's seemingly simple. That is because we'll use the neck area to convert from the face's higher resolution to the torso's lower resolution. So, let's get started.

First off, we'll use the bottom of the chin and the throat area to reduce the resolution, starting by tracing the new resolution, extruding the jawline loop downward, and contouring the area with the new target resolution:

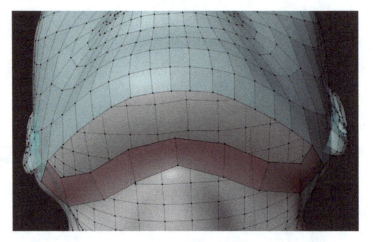

Figure 9.56 – New target resolution applied under the chin area

Now, we can make the necessary conversions, using the space between the new extruded faces to reduce the resolution from the chin to the neck:

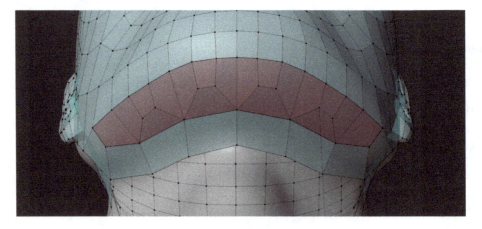

Figure 9.57 – Resolution decreased from the chin to the neck

As you can see, every three faces on the chin were turned into one toward the neck. How this looks will likely change from model to model, depending on the overall shape and resolution of the face, along with the resolution needed/wanted for the neck. Some models may even not need that reduction of the mesh resolution.

Keep in mind that this may not be necessary, if you're OK with the number of faces currently on the chin. If that's the case, we can draw the previous contour using the current resolution and fill up the area normally, which will certainly be easier.

We'll do the same thing for the back of the neck, except this time, instead of adding new geometry, we'll reduce what we have now on the upper neck:

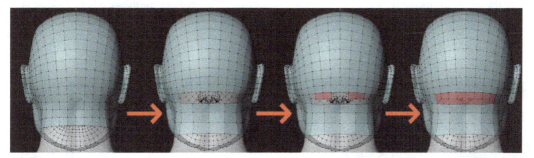

Figure 9.58 – Reduced resolution at the back of the neck

Perfect. Now that we have successfully reduced the resolution on the neck, we can finish retopologizing it. It's extremely simple due to its mostly cylindrical shape, so we can just extrude downward all of the faces currently going around the upper area of the neck, going for larger faces as we reach the shoulders, back, and chest. You can get as fancy as you want with the geometry, as there are many ways that will give decent results, but we'll stick to a simpler topology for our base meshes.

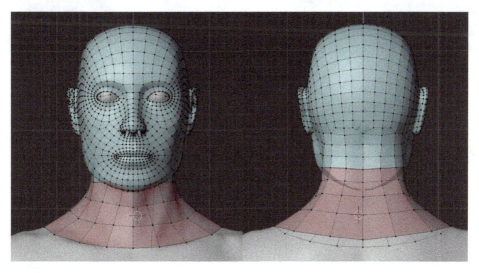

Figure 9.59 – Neck geometry

The neck geometry can look the same way for both base meshes. Now, onto the torso.

Torso

The chest is the biggest thing we'll have to retopologize, and like the face, it also has more than one final result that looks decent. In our case, we'll keep it relatively simple, with nothing too fancy, in order to make the process easier to understand.

Since the shape of the chest varies a lot from the male to the female base mesh, we'll have to look at them separately, starting with the male.

We'll add the most important loops first, which will initially be disconnected from the geometry we already have. The first loop goes around the shoulders:

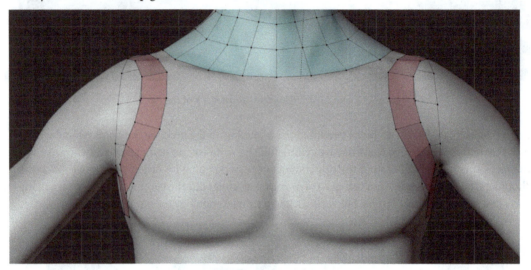

Figure 9.60 – Loop added around the shoulders

Notice how this loop curves slightly, following the shape of the muscles in the shoulders.

The next one is not quite a loop since it doesn't meet itself, but it goes along the edge of the pectoral muscles, contouring the lower area:

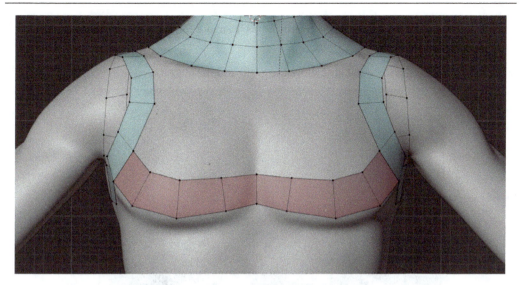

Figure 9.61 – String of faces added around the pectoral muscles

The next loop goes right under the one on the pectoral muscles, going around the entire torso and the back:

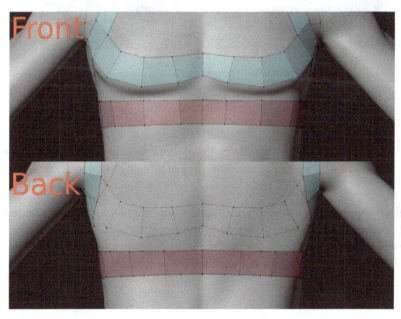

Figure 9.62 – Loop added under the pectoral muscles

Keep in mind the number of faces in the nearby loops, to make connecting these loops easier later on. In our case, this loop has 18 faces.

The last loop goes around the waist, just above the groin and the glutes:

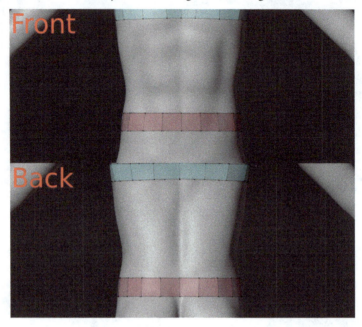

Figure 9.63 – Loop added at the waist

This loop has the same number of faces as the previous one, which makes connecting them extremely easy, and is what we'll start with.

When you're in a situation like this, **Grid Fill** won't work since these loops don't have any connecting edges. How nice would it be if we had an automatic way of bridging loops with the same number of faces together…

The **Bridge Edge Loops** function can be used through the **Edge** menu, which can be accessed using the *Ctrl + E* shortcut when two edge loops are selected (in our case the edge loops that, when connected, will cover most of the abdomen). Upon selecting the loops and pressing *Ctrl + E*, you'll be presented with a menu, from which we'll select the **Bridge Edge Loops** function:

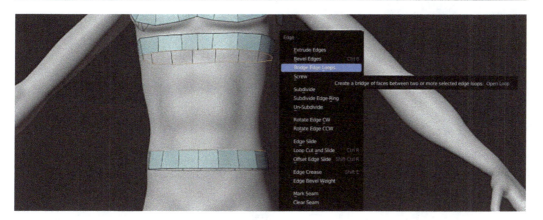

Figure 9.64 – Bridge Edge Loops function

Now, you'll see that the loops were indeed connected, but with only one face, meaning that there are no horizontal cuts. You can either add the necessary cuts using *Ctrl + R* to add more edge loops or adjust the **Number of Cuts** parameter on the settings menu that pops up in the bottom-left corner of the Viewport right after we bridge the edge loops together:

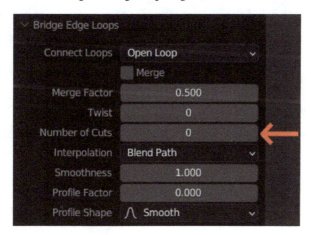

Figure 9.65 – Number of Cuts setting for the bridged loops

In our case, we used four cuts, but this may vary depending on the distance between the edge loops you bridged and the resolution you're aiming for. Either way, we should aim for faces that are as square as possible:

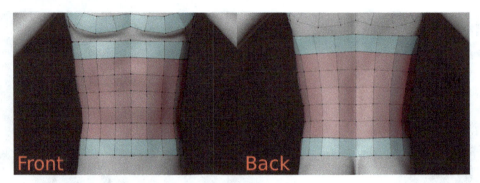

Figure 9.66 – Bridged edge loops on the abdomen

Now, we can go back to the upper area and begin connecting the rest of the loops, starting with the loop on the edge of the pectoral muscles and the one in the upper abdomen. This one should be relatively easy, especially if you kept count of the faces on each loop. We can extend this connection to the back too:

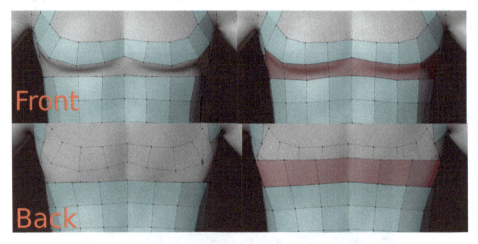

Figure 9.67 – Loops on the upper abdomen and chest connected

Now, before we fill up the area over the pectoral muscles, we'll connect the neck geometry to part of the shoulder loop, in order to have one empty space on the chest and one at the back, which we can handle separately. If you kept count of these faces, it should be extremely easy:

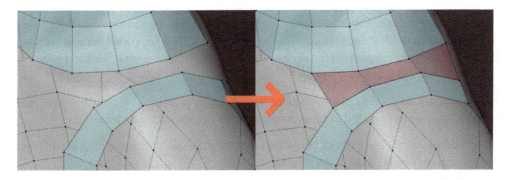

Figure 9.68 – Neck and shoulder connected

Now, we can fill up the space on the chest with a grid of faces, connecting the neck, shoulder, and chest loops. Don't be afraid to add more geometry if you need to, but avoid adding more loops to geometry that is already settled, such as the neck and face:

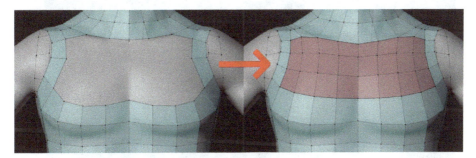

Figure 9.69 – Chest filled

As you can see, we added another loop to the connection between the neck and the shoulder and split one of the faces on the shoulder loops using *Ctrl + R*.

For the back, we'll use a similar logic, using a grid to fill out the empty space:

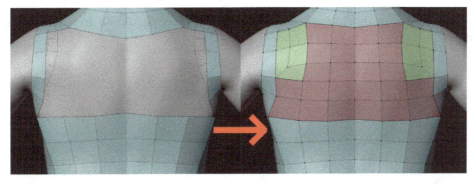

Figure 9.70 – Upper back filled

Now, you may have noticed that in the before image on the left, the top portion (shoulder/base of the neck) has four faces while the bottom of the space has three. That happened because we added another loop to the connection from the neck to the shoulder. To counter that, we had to use a 2:1 face conversion, marked in green.

These faces were not smoothed out too much for the sake of visualization, but now that we're ready to fill up the space, we can proceed to smooth out all of this new geometry we added.

Now we're finished with the male torso, so let's have a look at the female torso and explain the difference between the topology on the male and female torsos.

The only difference between the male and female torsos in terms of topology is the chest area, due to the breasts. For that reason, we'll start the female retopology with the shoulders and abdomen already filled up like the male base mesh, leaving only the chest and back with an empty space:

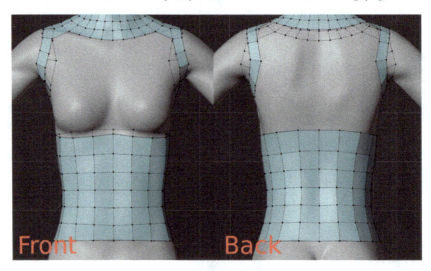

Figure 9.71 – Female torso with the chest and upper back left empty

The first thing we'll do is add a loop around each breast, again keeping in mind the number of faces of the nearby geometry:

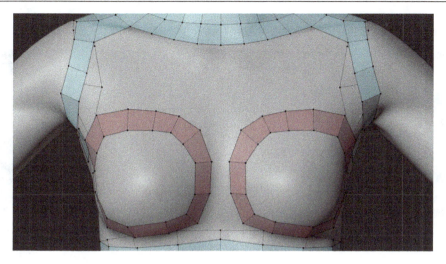

Figure 9.72 – Loop added around the breasts

Now we can use the **Grid Fill** function to fill the inside of these loops. Remember that if the number of faces in the loop is odd, **Grid Fill** will not work.

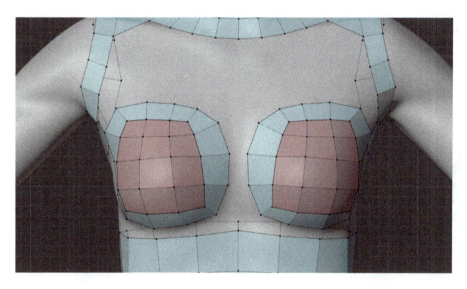

Figure 9.73 – Breasts filled up with Grid Fill

Now, we can begin filling the space under the breasts, extending the connection to the back:

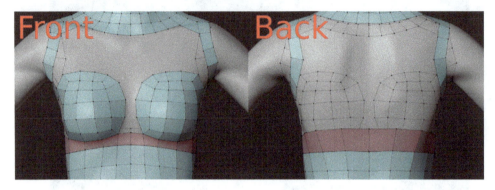

Figure 9.74 – Breast loops and upper abdomen loop connected

Notice how this loop follows the curvature of the breasts and has two faces in the middle, which we'll use to avoid unnecessary poles. You might have to add another vertical edge loop to the abdomen to be able to leave those two faces.

Now, we can connect the two middle faces on the base of the neck to those faces between the breasts:

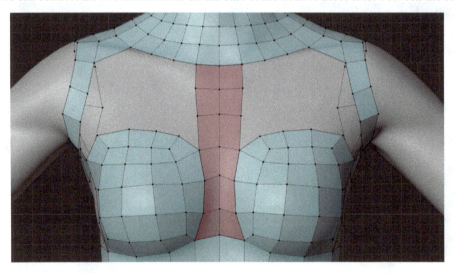

Figure 9.75 – Loops on the base of the neck and under the breasts connected

Perfect. Now we can fill up the remaining spaces with a grid, similar to what we did with the male base mesh:

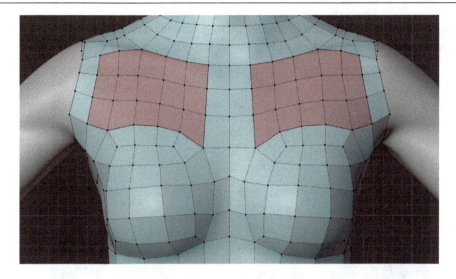

Figure 9.76 – Empty space on the chest filled

Again, similarly to when we did the male topology, we had to add one loop to the connection between the neck and the shoulders and one more to the shoulder loop itself.

Now, we can extrude the faces on the sides of the breasts to the middle of the back, following the number of faces we currently have on the back:

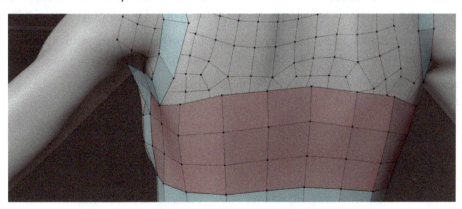

Figure 9.77 – Faces on the sides of the breasts extruded

Now we're left with a very similar situation to the male base mesh on the upper back, where we have to fill the space while reducing the resolution from the top to the bottom part:

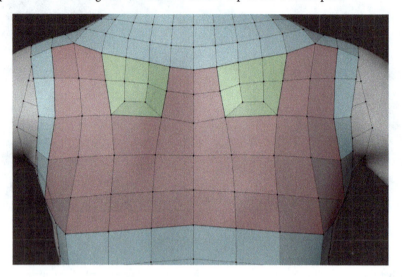

Figure 9.78 – Upper back filled

Notice that, differently from the male base mesh, the reduction in resolution happened toward the middle of the back rather than near the shoulders. Note, however, that this could have happened in the male base mesh too, depending on the resolution of the neck and how the topology is laid out.

Perfect. We have successfully finished retopologizing the torso of both our base meshes. Now we can move on to the arms and legs, since they share a very similar logic.

We will also return to only showing the process for the male base mesh, since from now on the topology between male and female isn't very different.

Here is the topology we have so far on both the male and female meshes:

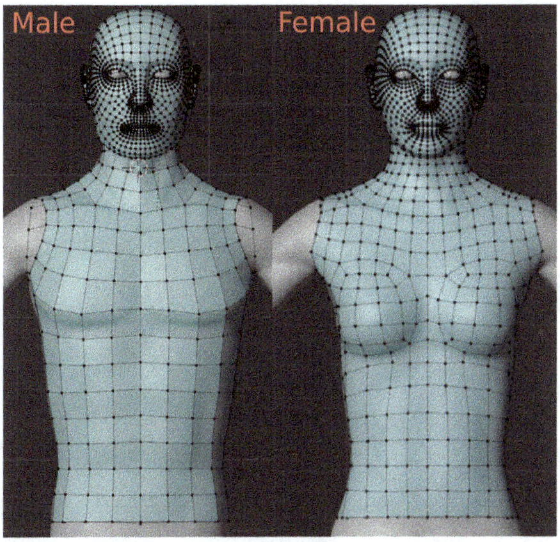

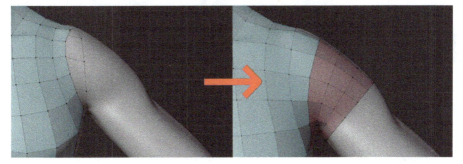

Figure 9.79 – Current topology

Now, back to the process, we'll proceed to retopologizing the arms.

Arms

The arms are actually extremely easy, and since they have a mostly cylindrical shape, there are a few tricks we can use to reduce the manual work. However, we'll begin by extruding the shoulder loop over the shoulder and armpit area manually, since the armpit is a tight space and sits right under the shoulder.

We'll make a few loops around the actual shoulder, keeping the faces on the armpit thinner due to the tightness of the area. Don't forget to smooth the geometry once you're done:

Figure 9.80 – Shoulder covered

We could continue extruding these loops until we reach the wrist, but there's a trick to make the entire arm retopologize itself in a way, leaving us with a minimal amount of manual work to do.

We will start by creating a vertex loop on the wrist with the same number of vertices on the shoulder loop. You can do that by either duplicating all the vertices on the shoulder loop with *Shift + D* or duplicating a single vertex and extruding it around. The important thing is that it must have the same number of vertices as the shoulder loop and go around the wrist. Also make sure that the vertices on the wrists and the shoulders are roughly alinged:

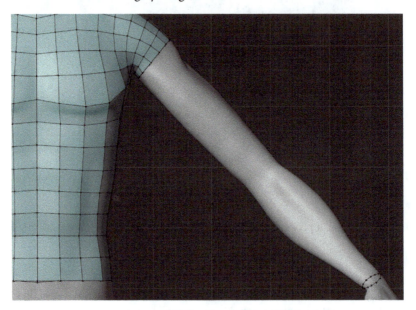

Figure 9.81 – Vertices around the wrist placed

Now, we can select the wrist vertices and the shoulder loop vertices and use the **Bridge Edge Loops** function, which will connect the loops. Finally, like we did with the abdomen, we can change the **Number of Cuts** parameter in the menu that pops up in the bottom left (we used 18 cuts) until we get square faces. The **Shrinkwrap** modifier will take care of maintaining the geometry on the surface:

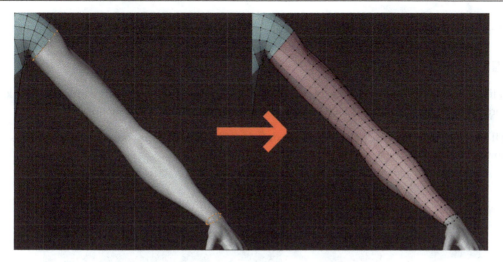

Figure 9.82 – Bridge Edge Loops function used in the arm

Now, you may notice that the geometry left by the function is a little crooked at some points. It may help to undo this action, duplicate the **Shrinkwrap** modifier, apply the top one, and do it again. Even then, some geometry might be slightly crooked, but 90% of the work on the arm was done for us. What's left is smoothing out the vertices. You should have something like this:

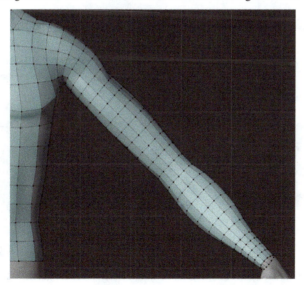

Figure 9.83 – Topology on the arm smoothed out

Feel free to add a few extra loops around the elbow, since it deforms more and would need slightly more resolution to deform to its full range of motion.

Perfect. Our arms are looking nice, so let's move on to the legs, since they follow a very, very similar logic.

Legs

Just like the arms, the legs are extremely easy to retopologize, but they also require a little bit of work before letting Blender do the rest, since we also have the groin area to worry about.

We'll start by extruding the two middle faces down, until we reach their counterparts at the back, following the shape of the groin:

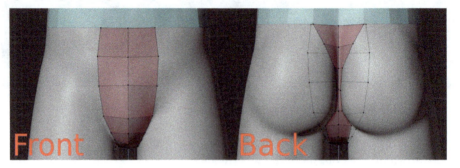

Figure 9.84 – Faces added to the groin

Now, from that same geometry, we'll extrude the faces horizontally around the waist, also covering the glutes:

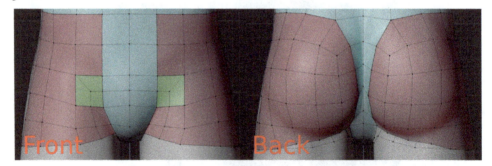

Figure 9.85 – New topology added to the waist

Notice how we had to increase the resolution from the abdomen to the groin/waist to better capture the curvature in the area. This hasn't smoothed out yet in order to be easy to see.

Now, we're ready to let Blender do the rest of the work.

Start by duplicating all of the vertices that make up the contour of the leg, including the ones at the bottom of the groin (we need a closed loop of vertices, like on the arm) and place them on the ankle:

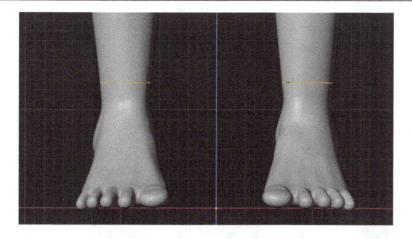

Figure 9.86 – Vertices placed on the ankle

Now, we can use the **Bridge Edge Loops** function again to retopologize the entire leg at once. Like with the arm, you might have to play around with the settings to see what works best. We used 23 cuts for this.

After smoothing out the geometry left by this operation, you should have something like this:

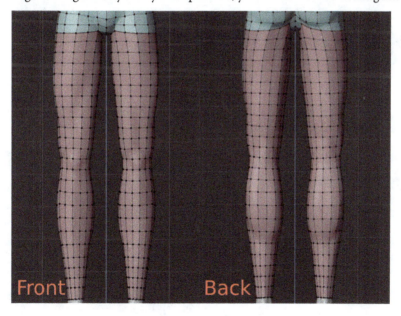

Figure 9.87 – Legs retopologized

Perfect, the legs are done.

Now, only the hands and feet are left, which can sometimes be trickier. Let's start with the feet since they are a little simpler to deal with.

Feet

Retopologizing the feet can be a little tricky, especially because of the toes, so we'll have to pay a little more attention to how we do things here. So, take your time when working on the foot topology, because things tend to get messy and confusing.

Let's start by defining some important loops. The first one extends from the edges of the ankle, going all around the sole and ending at the opposite ankle:

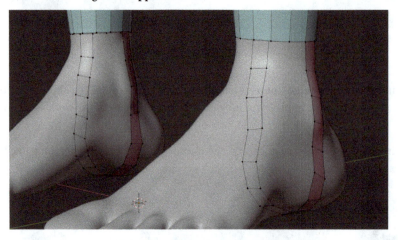

Figure 9.88 – String of faces added under the sole

The next loop extends from this one, going back toward the heel all the way to the other side of the foot:

Figure 9.89 – Faces added to the heel

For now, let's not worry too much about the loops being a little crooked or the number of faces not matching between the loops. We can always smooth or add and/or remove geometry later when we're connecting these loops.

The next two loops we're going to add are one around the front of the foot, near the toes, and one around the base of each toe:

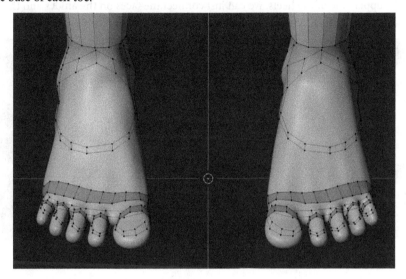

Figure 9.90 – Loops added around the toes and near the front of the foot

Notice how the face placement on each toe is roughly matching that on the front of the foot. This will make connecting these loops later on a little easier. It is extremely important to keep the number of faces in each toe as an even number, such as 6, 8, 10, 12, and so on.

Now, the last loop goes around the heel itself, in a circle:

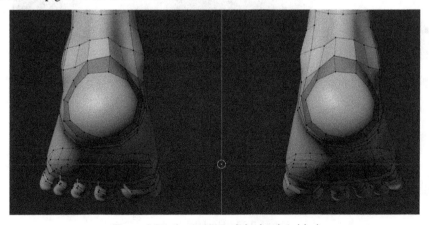

Figure 9.91 – Loop around the heels added

As you can see, this loop can be connected to the string of faces we added on the back of the heel earlier. Make sure to try and match the number of faces in this loop to the loop that goes under the sole, since it will make filling up the space much easier.

Perfect, now we can start filling up the holes.

Starting with the upper side of the heels, we can just connect the faces on the ankles and the faces we added at the back of the heel, adding or removing geometry as we need:

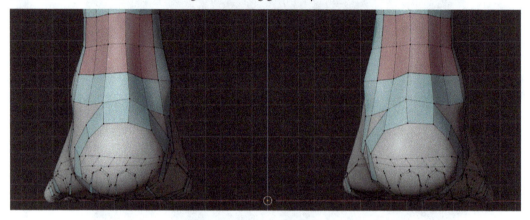

Figure 9.92 – Hole on the heels filled

Now, let's finish the heel by using the **Grid Fill** function from the **Fill** menu (*Ctrl + F*) to fill the circle loop (you can add/remove faces to make filling it with a grid possible; just make sure to add those changes to the nearby loops as well if you matched the number of nearby faces in those loops). Then, we can fill the surrounding space normally, which, again, will be easier if you kept count of these two loops. You should have something like this:

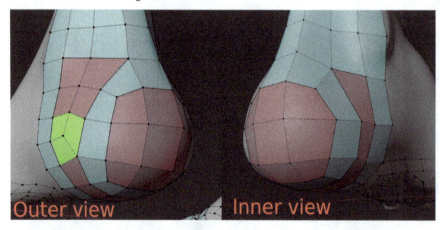

Figure 9.93 – Heel finished

As you can see, we needed to change the resolution one time for the outer side of the foot.

Perfect. Let's start working on the middle of the foot.

First, we'll connect the existing geometry on the tip of the foot with a string of faces on the top and one on the sole:

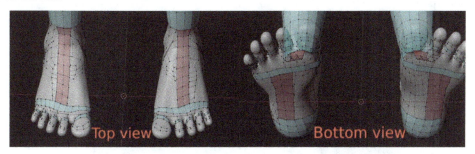

Figure 9.94 – Ankle connected to the front of the foot

These strings don't necessarily have to be exactly in the middle, since the foot isn't symmetric itself.

Now, we can fill up the spaces left, trying to use a grid-like geometry, which may not be easy due to the distance between this loop and the rest of the geometry. Don't worry, though, as we can easily add/remove geometry, since the loop on the front of the foot is still disconnected from the rest.

Let's start by filling up the outer side, connecting each face from the front to its counterpart near the ankle/heel:

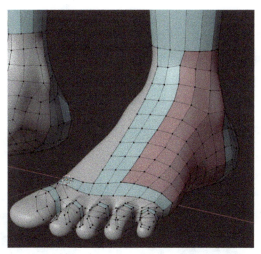

Figure 9.95 – Outer part of the foot filled in

It can help to go around the hole, filling a bit of each edge at a time, narrowing the hole more and more. Using references here can be extremely helpful to understand how the topology flows on complex shapes like this.

Now, we'll fill the inner side, following a similar logic of connecting the faces on each loop with their counterparts:

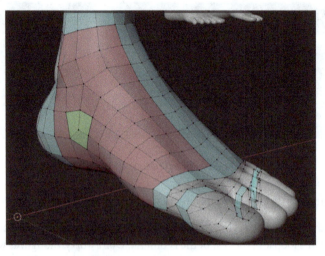

Figure 9.96 – Inner side of the foot filled in

In that case, we had to change the resolution to accommodate the number of faces in the different loops.

Now, let's do the tricky part, which is connecting the base of the toes to the rest of the geometry. If you made them all with the same number of faces, it may be easier.

We'll start by connecting all the toes together from the sides:

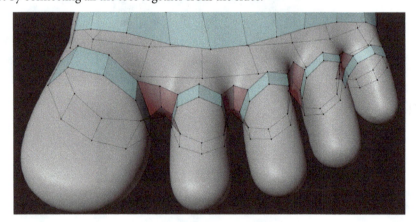

Figure 9.97 – Toes connected

Each toe is connected to the next using the two middle faces from the loop.

Now, connect each of these faces with the next one. Do this on the underside of the toes too to finish the second loop around each toe:

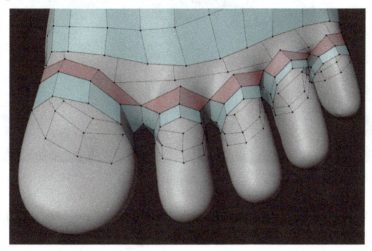

Figure 9.98 – Second loop around each toe finished

Now, we can connect the top side of the toes with the rest of the body:

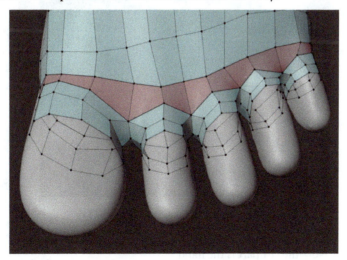

Figure 9.99 – Toes connected to the foot

If you need to, you can add more edge loops to the toes or the foot with *Ctrl + R* to make it possible to connect them in a clean way. Just don't forget to smooth out the new geometry afterward.

Now, to the front of the sole. It can be a little trickier since the hole left is bigger than what was on the top, but we can get around that by changing the resolution to allow the faces to be connected:

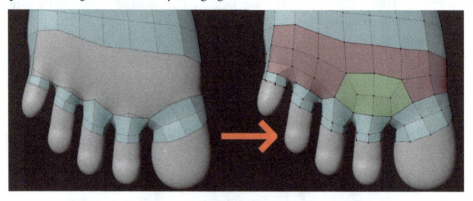

Figure 9.100 – Hole on the sole filled up

Now that the tricky part is done, we can finish the toes by extruding the ring of faces to the tip of each toe and, when possible, using **Grid Fill** to fill up the tip:

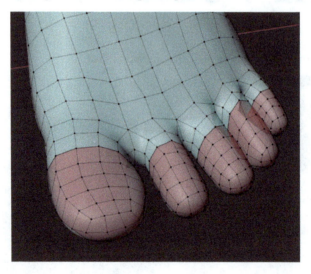

Figure 9.101 – Toes filled

Alright, looking nice. Now, the last part is the hand.

Hands

The hands can be just as tricky to deal with as the feet, but we can deal with it in a similar way, especially with the fingers.

Let's start by blocking out some loops. We'll make one loop around each finger, including the thumb. Remember to keep the number of faces in each loop an even number so as to not have trouble with the tips of the fingers later on:

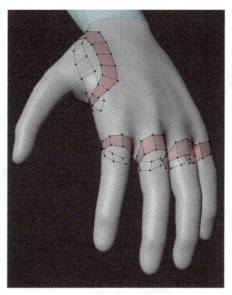

Figure 9.102 – Loop added around each of the fingers

Now, we'll connect all the fingers, using the sides of the fingers as the connecting points:

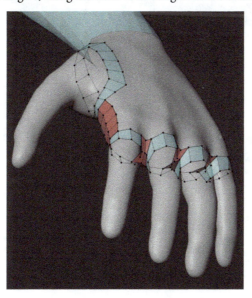

Figure 9.103 – Fingers connected

Notice how there are only two faces connecting each finger (except for the thumb, of course). This will make it easier to add another loop near that joint of the fingers, increasing the range of motion of that finger when and if we need to move it.

Now, let's finish retopologizing the fingers, remembering to close the tips so that if we need to add/remove loops later on, the number of faces remains even throughout the whole process:

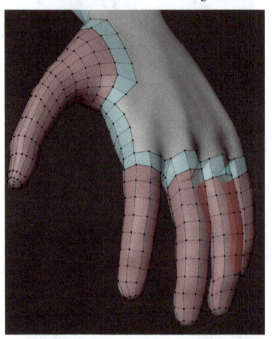

Figure 9.104 – Fingers retopologized

Remember that if you have an even number of faces around the fingers, you will be able to use the **Grid Fill** function to close the tip automatically.

Now, we'll add another loop at the base of the fingers (both at the back of the hand and at the palm side), using the connection between the fingers as part of the loop:

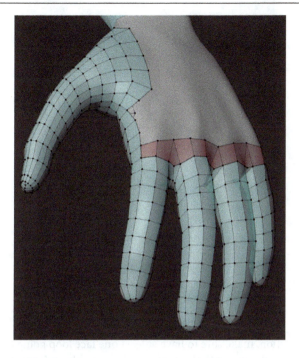

Figure 9.105 – Additional loop added at the base of the fingers

Notice how there's a diamond-shaped quad in the middle of the fingers. This both reduces the resolution from the fingers to the palm and makes it so that any edge loop added there goes downward and around the fingers. The same diamond trick was used on the underside of the fingers, making a complete loop at the base of each finger. It's basically the same as a 2:1 face conversion.

Now, let's connect the sides of the pinky finger to the wrist so we can work on the back of the hand and the palm separately. Make sure to have the same number of faces as on the other end of the hand (although you can add or remove geometry to it later):

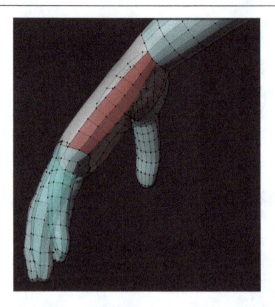

Figure 9.106 – Pinky finger connected to the wrist

While making this connection, make sure to have at least one face loop going around all the fingers except the thumb, which we'll connect in the next step. You can check for that aspect by using *Ctrl + R* and adding an edge loop to see whether it goes across all the fingers (highlighted in yellow). If this doesn't happen, you may need to change how the topology on the tip of the fingers is laid out:

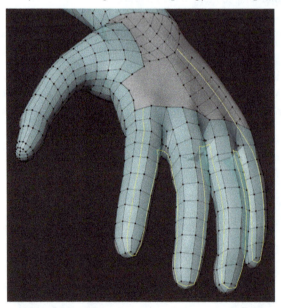

Figure 9.107 – Edge loop going across all the fingers

In our case, the edge loop stops at the index finger. We'll have to connect that face to the thumb and the thumb to the wrist following that same loop to ensure that there is at least one loop going across all five fingers:

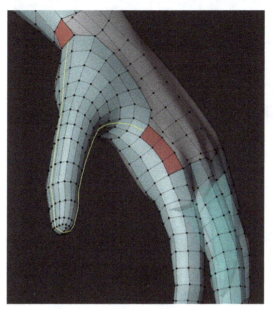

Figure 9.108 – Loop across all five fingers completed

See how the edge loop highlighted in yellow follows the contour of the thumb horizontally, completing the loop we tested out in the previous step.

Perfect. Now we can start filling in the holes left on the palm and the back of the hand. We'll start with the back of the hand. These parts can be tricky to get right, since the number of faces in the knuckles may not be the same as the number in the wrist. To fill those holes, we might need to add more edge loops to the wrist or knuckle with *Ctrl + R*. If you need to add/remove loops from the wrist, though, remember that it is connected to the rest of the body, so you will not only affect the arms. When that's the case, we recommend aiming to only add or remove geometry to loops that meet the symmetry line on either the chest or the back, so as to avoid changing the geometry on the face, legs, and/or feet. Nevertheless, we'll aim for a grid-like surface when filling these areas. Here's the geometry we got on the back of the hand:

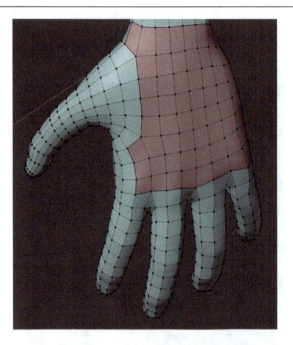

Figure 9.109 – Back of the hand filled

Nice, looking good. Now, we'll apply the same logic to the palm:

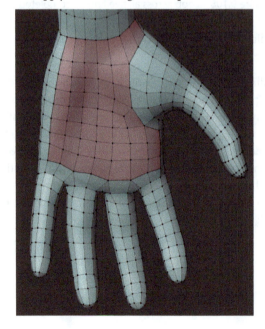

Figure 9.110 – Palm filled in

Again, we aimed for a grid-like surface. You can get way more complex geometry that captures more detail, but for learning purposes, we chose a simple yet effective geometry that can be subdivided and deformed properly.

Congratulations! We are finished with the retopology. Here's a full-body view of how the new topology looks overall:

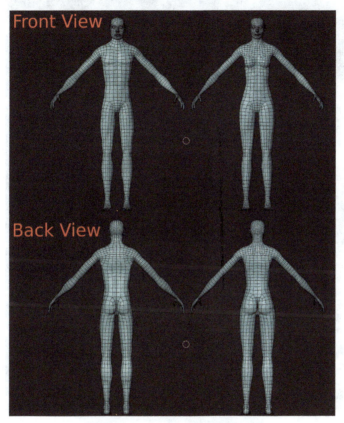

Figure 9.111 – Full-body topology

Perfect. Now we have much better topology and way fewer faces on our models, with only around 6,000 faces on each.

If we return the mesh to how it originally looked before we started retopologizing, though (by disabling the **In Front** parameter, changing **Color** back to white, and isolating the new mesh with /), you'll see that we lost a lot of detail due to the very small number of faces. We can get all of that detail back by applying all the modifiers that are currently in our meshes, then adding a **Multiresolution** modifier followed by a **Shrinkwrap** modifier with the target set as the high poly, not the retopologized, version (this will increase the number of faces, though, so if you don't want that, you'll have to consider baking a normal map, as we covered in *Chapter 5*). Here's what your modifier stack should look like:

Figure 9.112 – Modifier stack for reprojecting the details

Now, as we increase the subdivisions on the **Multiresolution** modifier, you should see all the detail come back since the resolution of the mesh is increasing, and afterward, it's being shrinkwrapped onto the original sculpt. Here's what different levels of subdivisions look like:

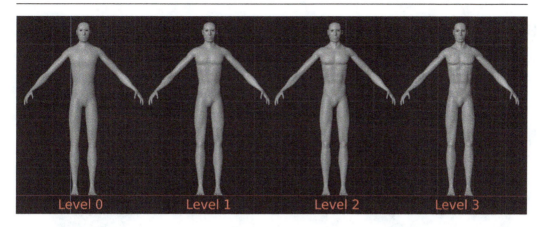

Figure 9.113 – Different subdivisons

Now, we'll subdivide the mesh until we capture all of the necessary detail, which in our case is level 3. Then, with our mesh still subdivided, we'll apply the **Shrinkwrap** modifier only, since we want the flexibility of going back and forth on levels of detail, at least inside Blender.

Perfect. Now we have a detailed mesh that has clean geometry and can be subdivided properly.

Well, what if we want to texture our character? Where would we put the seams to UV unwrap it? This may not be so intuitive in a mesh like this.

UV unwrapping our base meshes

UV unwrapping a human can sound a little daunting at first, but most times it's simpler than it seems.

Again, just like most parts of the process, there are many variations that can work just fine. In our case, we'll go for a simpler unwrap. It's also worth noting that the same unwrap works for both male and female bodies, given that they have relatively similar geometry.

We'll start with the neck and head and go down. At the end of each section, we'll have a picture of what the UV islands for that part look like. You don't need to unwrap the parts separately, though; this is just for demonstration.

Head and neck

We'll unwrap the head and the neck together in a single UV island. For that, we'll only need two seams. The first one will be placed at the base of the neck:

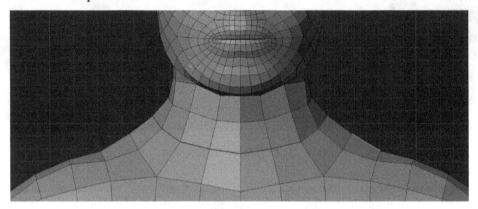

Figure 9.114 – Seam added at the base of the neck

The second loop goes on the back of the head, starting at the base of the neck and going up until the middle of the forehead, right along the line of symmetry:

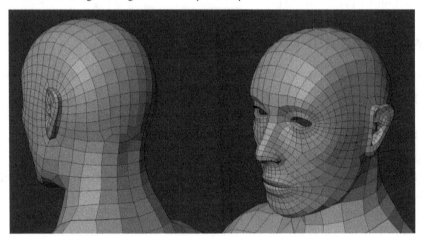

Figure 9.115 – Seam placed on the middle of the head

Additionally, you could add another seam going from the base of the neck until the chin, also at the line of symmetry, but that's optional. If we unwrap only the head, you should see this:

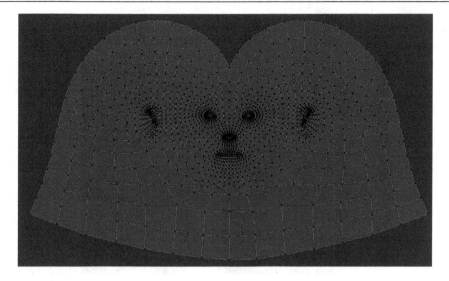

Figure 9.116 – Head UV island

This head unwrap is extremely common and allows for texturing just fine. It's also possible to place seams around the ears, separating them from the head if necessary, for more resolution in that area while texturing.

Now, for the torso.

Torso

The seams on the torso are just as simple, since we'll separate the front and the back.

Let's add a seam to each of the shoulders and another loop from the base of the neck to that shoulder seam:

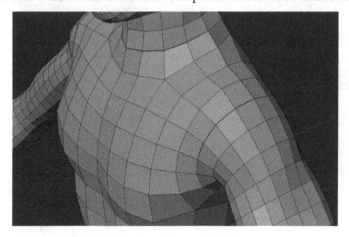

Figure 9.117 – Seam added on the shoulders and trapezius

Now, we'll add seams around the top of each leg and connect those seams with another seam going through the groin:

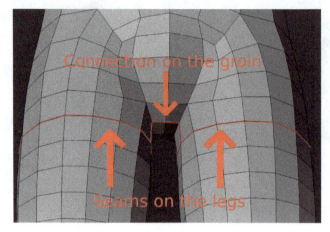

Figure 9.118 – Seams on the legs and groin added

Now, the last loop is on the sides of the torso and will connect the shoulder seam to the leg seam:

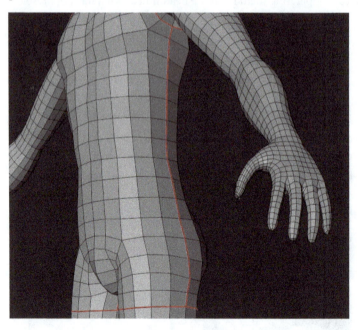

Figure 9.119 – Seam on the sides of the torso added

Now, we're finished with the torso, which has two separate UV islands. They should look like this:

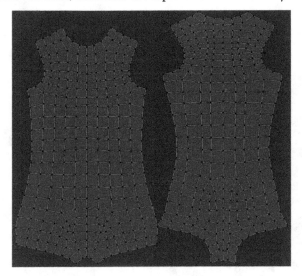

Figure 9.120 – Torso UV islands

Perfect, the torso is looking good. Now, let's make the arms.

Arms

Just like retopologizing, unwrapping the arms is super easy, as they mostly resemble a cylinder.

We'll only need to add one seam at the wrist and connect it to the seam on the shoulder:

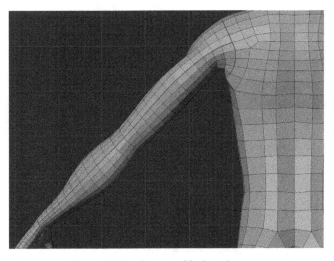

Figure 9.121 – Seams added on the arms

The seam connecting the wrist to the shoulder was put on the back to better hide potentially visible seams on the textures. Here's what the UV island looks like:

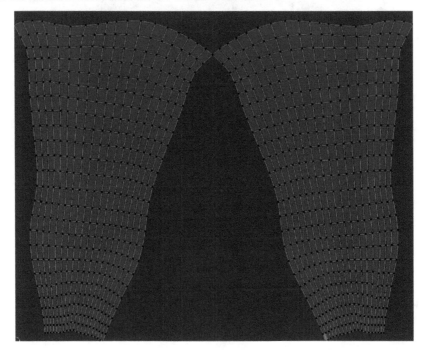

Figure 9.122 – UV islands for the arms

Note that these are the islands for the two arms, so one island per arm.

Now, let's unwrap the hands.

Hands

Surprisingly enough, the hands are extremely simple to unwrap, given you have decent topology on them. In fact, we'll only need one seam.

The seam we'll place on the hands will be put across all the fingers, including the thumb. Sound familiar? That's why we insisted on having at least one face loop going around all the fingers:

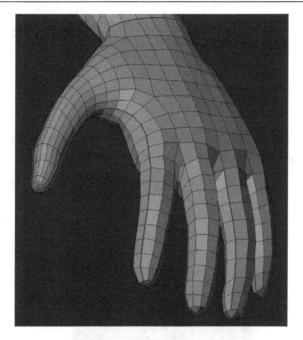

Figure 9.123 – Seam placed on the hand

Perfect. Now each hand also has two separate UV islands, which look like this:

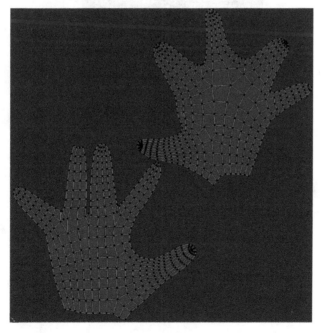

Figure 9.124 – Hand UV islands

Note that some degree of distortion is to be expected, as the shape of the hand is complex, but as long as the unwrap looks somewhat proportional, you should be fine.

Let's move on to the legs.

Legs

The legs follow the same logic as the arms, so we'll add one loop to the ankle and connect that to the loop we added to the upper leg when placing the seams for the torso:

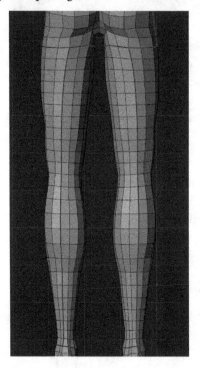

Figure 9.125 – Seams added on the legs

The UVs should look similar to the arms as well.

Finally, let's unwrap the feet.

Feet

Again, unwrapping the feet is much simpler than making or retopologizing them. We'll only need two seams.

The first one goes across all toes, just like with the hands:

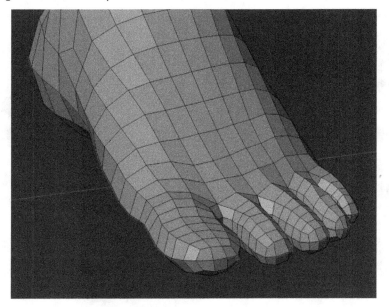

Figure 9.126 – Seam added across the toes

Now, the last seam we need goes on the heel, from the ankle seam until the seam we just added. This splits the back part of the foot into two parts and eliminates most of the stretching that could occur in that area:

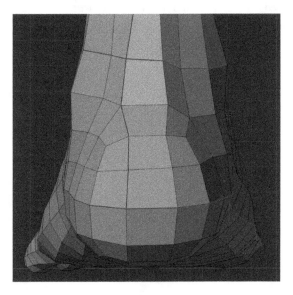

Figure 9.127 – Seam added in the heel

Now, if we look at the islands for the foot, we should see this:

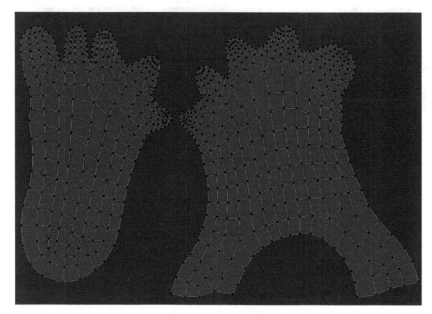

Figure 9.128 – UV islands for the foot

The island for the upper side of the foot is more distorted due to the seam we added at the end, to split the heel into two parts. These islands might look weird, but there's minimal stretching.

Now, with every seam placed, we can hit *U* and select **Unwrap** from the menu that appears. This will unwrap and pack the geometry for the full body, and it should look something like this straight out of the box:

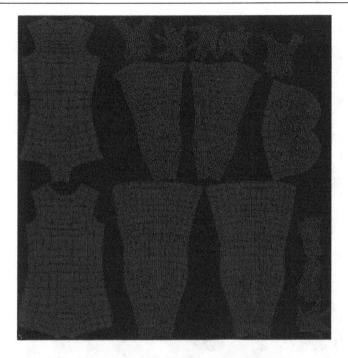

Figure 9.129 – Blender's pack for our UVs

It's a good starting point, but we can do better. Notice how some of the islands are not grouped near their counterparts, such as hands and feet. Also, see how the face is small compared to the rest. Relatively, the head is smaller, yes, but for texturing, we usually leave more room for the head and face, since it's one of the most important parts of any character (sometimes even leaving a whole UV map just for the head, in what's called a UDIM tile, but this won't be covered). So, we'll have to manually fix those two mistakes, and most importantly, try to occupy as much space as possible while keeping the islands as straight as we can. Since most unwraps look different, we can't cover the exact process of doing so, but generally, we keep these aspects in mind.

We ended up with something like this:

Figure 9.130 – Manual pack for the full human body

Now that's looking better. All of the similar parts are grouped together and the space is being occupied more efficiently than before. This will make it so that not only will our texture look better, but we'll also be able to modify and identify the different parts more easily. We won't cover how to texture a character, though, since we've already covered the basics of texturing.

Summary

In this chapter, we covered in depth how to properly retopologize a humanoid character, some tips and tricks that can even be used in other types of characters too, with detailed explanations on how to set up the retopology mesh with better visibility, as well as why the geometry needs to be laid out in a specific way, making use of essential loops to guide the retopology process. At the end, we also covered how to prepare our mesh for texturing by efficiently UV unwrapping and packing the new geometry.

Now we have a fully optimized base mesh that we can use for multiple purposes, but it's stuck in that position, so how do we move it? How do we add the proper joints to the right places to add different poses? We'll cover that in the next chapter.

10
Rigging the Base Meshes

After we completed the retopology on our character, we'll need to add a **rig** to it if we ever want to pose and/or animate it in any way. A rig is basically a skeleton we add to a model that can move several vertices at once and is usually used to mimic how joints move in real life. Even though it's mostly known for bringing characters to life, a rig can also be used on inanimate objects that need moving parts. This chapter is essential if you ever plan on having a character that can be posed and animated at some point.

This chapter will cover the following:

- The different types of movable joints in the human body
- Introduction to the rigging system in Blender
- **Forward Kinematics (FK)** rigging
- **Inverse Kinematics (IK)** rigging
- How to automatically generate a full rig with the built-in add-on Rigify

Before we start any rigging work, though, we'll have to understand the different types of joints in the human body and how they will affect our rigging.

Types of movable joints in the human body

The human body is capable of moving in a vast variety of ways, which can be somewhat overwhelming to some beginners that aim to make functional rigs in the future, which sometimes tend to not take into account the limitations of some major joints in the body. Yes, some people are more flexible than others, but the joints have their limitations regardless, and when it comes to a rig that is supposed to imitate the way that a human moves to a more realistic degree, those limitations can be the difference between a character that looks believable in different poses and one whose limbs seem to be broken in some way. So, let's have a look at some of the main types of joints and how they move, and for each joint, there will be a simplified model to demonstrate both its shape and how it moves.

Note that we'll only cover the joints that have a bigger range of motion, as they are the only ones we really have to focus on while rigging a character. Those belong to a group called **Synovial Joints**, if you need to do any further research down the road. Let's start with the most movable ones and move down to the least movable.

Ball-and-socket joints

This joint is just what it sounds like: it's a ball inside of a socket and allows for pretty much any movement along any axis, such as flexion, extension, abduction, adduction, rotation, and circumduction (see *Figure 10.1*). In the human body, we only have two of those joints – one at the shoulders and the other at the hips.

The shoulder joint has a shallow socket, which grants a wider range of motion while having less stability. The hip joint, on the other hand, has a deeper socket, which makes it more stable but limits its range of motion.

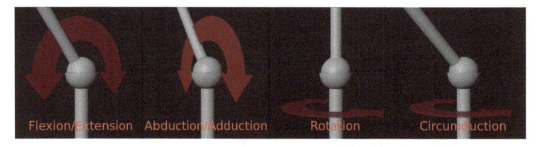

Figure 10.1 – Movements of a ball-and-socket joint

Ellipsoid joint

An ellipsoid joint works very similarly to a ball-and-socket joint; however, due to its oval shape and the ligaments around it, this joint is unable to rotate.

It still allows movement along two axes, meaning it's still able to do flexion, extension, abduction, adduction, and circumduction:

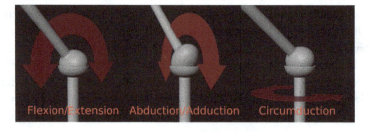

Figure 10.2 – Movements of an ellipsoid joint

When abducting and/or adducting, the joint slides around its socket toward the center. Here's a demonstration of how it happens:

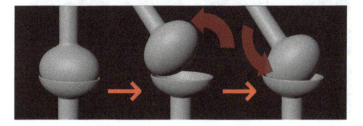

Figure 10.3 – Ellipsoid joint sliding into the socket

A good example of this type of joint is the wrist, in which the group of bones there rotate inside the socket of the radius (one of the forearm bones).

Next, there's the saddle joint.

Saddle joints

A saddle joint is similar to the ellipsoid joint, with the rotation also being extremely limited, mostly due to its shape, which looks like this:

Figure 10.4 – Saddle joint

See how the parts connect at a 90° angle relative to each other, preventing most, if any, actual rotation from happening.

Even though the shape of this joint is very different, it still allows for flexion, extension, abduction, adduction (see *Figure 10.5*), and circumduction (see *Figure 10.6*). Some of these may be harder to do in one smooth movement due to the shape of the joint, though (especially circumduction). This makes the movements of this joint look slightly stiff.

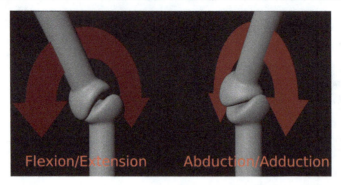

Figure 10.5 – Flexion, extension, abduction, and adduction on a saddle joint

Notice how the joint glides around the other part during abduction/adduction.

Now, the circumduction of this joint happens in an interesting way. The moving part slides around the other from one side to another, as well as flexing and/or extending. Here's what a full circumduction looks like for this joint:

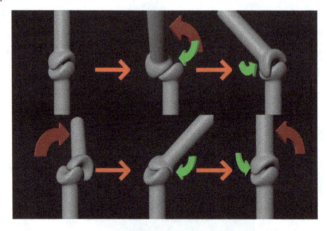

Figure 10.6 – Full circumduction on a saddle joint

An example of where this joint is located in the body is at the first joint of the thumb (near the wrist) and the sternoclavicular joint, which is the joint between the clavicle and the chest bone at the base of the neck.

Now we'll start getting into the less movable joints, starting with the hinge joint.

Hinge joints

This is a very simple joint, only allowing movements along one axis. Due to its shape, we can only flex and extend this joint, but the movement it provides is very stable:

Figure 10.7 – Movements of a hinge joint

The perfect examples of this type of joint in the human body are the elbows and knees, which can only be flexed or extended. We can only twist these limbs because of the joints on the shoulders and the hips.

But wait, we can twist our hands without moving our shoulders, so how can we do that? Well, the next joint is the answer.

Pivot joints

This joint also allows for just one movement, and that is rotation. A bone with a cylindrical shape fits into a ring of bone and ligaments, around which it can rotate. It looks similar to this:

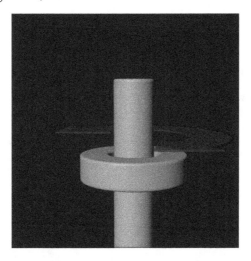

Figure 10.8 – Movement of a pivot joint

A good example of this type of joint in the human body is the one that is located just below the elbow, responsible for twisting our forearm and our hand as a consequence.

This happens because there's another joint near the elbow, the radioulnar joint, which rotates the radius (one of the bones in the forearm). Here's a simplified version of that:

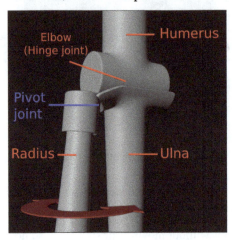

Figure 10.9 – Simplified radioulnar joint

Notice that the radius fits partially into a ring of bone on the ulna. The rest of the ring is made of ligaments, which allow rotation while keeping the bone in place.

When we rotate our wrist, the base of the radius (near the wrist) rolls over and around the ulna, making the radius cross over the ulna.

Now, onto the last synovial joint: the plane joint.

Plane joints

The plane joint is not as intricate as the others, and it's usually located in areas where there's not much movement.

The plane joint is also very similar to how it sounds. It's made of two relatively flat surfaces that can glide or rotate around:

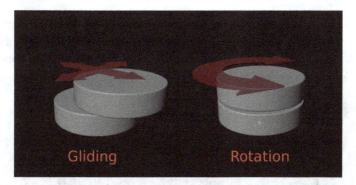

Figure 10.10 – Movements of a plane joint

This type of joint is usually seen in groups, such as the carpal bones in the hands (in the palm, near the wrist) and the tarsals of the foot (in the middle and back of the foot). They don't have much movement since they are held in place by ligaments, but they do allow for some gliding and rotation.

Another example is the acromioclavicular joint, which is the joint between the clavicle and the scapula (the shoulder blade). When we raise our shoulders, for example, the angle of this joint adjusts to keep the scapula in a vertical position.

Now that we're finished with the synovial joints, we can finally start rigging our characters.

Rigging our character

Rigging is the process of adding a digital "skeleton" to our mesh, in order to move and pose it more easily. This process is commonly used to pose/animate characters and creatures, but we can rig machines and inanimate objects too.

For our purposes, though, we'll focus on making a simple rig for humanoid characters, using the base meshes we made and optimized in the previous chapters. Note that this chapter will not include face rigging, since it can become very complex very easily. However, also keep in mind the geometry we have on the face works great for face rigging in case you want to add it later.

Alright, so how do we add a bone to the scene? Simple, they are on the **Add** menu (*Shift + A* in **Object Mode**), just like every other object. The difference is that, instead of adding a mesh, we'll add an **Armature**. So select the **Armature** option from the **Add** menu, and add a **Single Bone**:

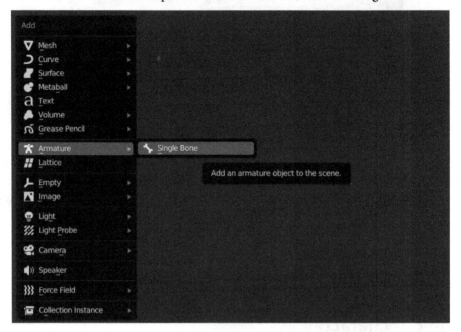

Figure 10.11 – Adding an Armature with a Single Bone to the scene

Now, upon selecting that option, an armature with a single bone should be added to your scene. And it looks like this:

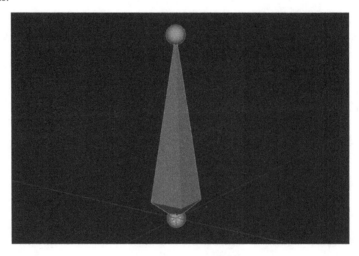

Figure 10.12 – Bone added

Keep in mind that these bones don't appear in the final render and are just a visual representation of the different joints that make up the rig so that we can move and manipulate them according to our needs.

Before we start positioning the bones, we'll have to make sure that they are seen from any angle, even if we put them inside the mesh (which we'll do). To do that, let's activate the **In Front** option as we did when retopologizing our base meshes to see what we were doing better.

With the bone selected in **Object Mode**, go to the **Object Properties** menu, find the **Viewport Display** submenu, and check the **In Front** option:

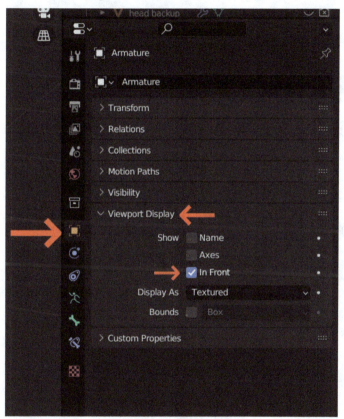

Figure 10.13 – Enabling the In Front option for our armature

Perfect, now we'll see the armature from anywhere.

In order to tweak the bone's size, we recommend using **Edit Mode** (most of the shortcuts used while editing a normal mesh still work with a bone, such as *G*, *S*, and *R*). This way, we can have better control over the size of each bone, since in **Edit Mode**, we can tweak the location of the **Root** and the **Tip** of the bones separately:

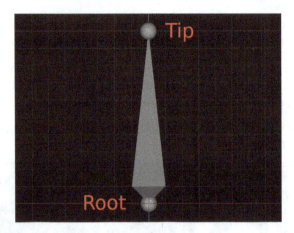

Figure 10.14 – Tip and Root of a bone

Keep in mind that the rotation happens using the root as a pivot point. Let's start by adding the bones to the spine.

Spine

For the spine, we'll move this bone we just added to the center of the waist, at the base of the back, since this is roughly at the height where our center of mass is:

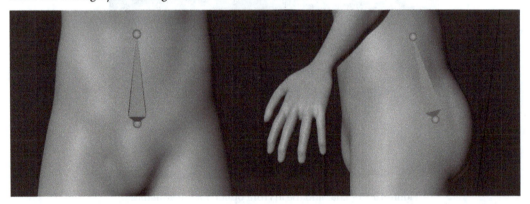

Figure 10.15 – Bone positioned at the waist

Remember that the first bone you add is the **Master Bone**, and it will move all of the other bones with it when it moves. That's why, in our case, we put this bone near the center of mass.

Now, we can make the rest of the spine by extruding the tip of this bone using *E*, following the general shape of the spine. Note that you don't need one bone for every single vertebra of the spine. Simplifying that to around 6 or 7 bones is enough. And remember to keep it centered and aligned with the line of symmetry since the spine is at the center of the body:

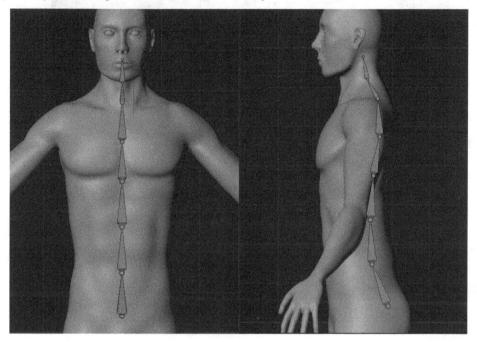

Figure 10.16 – Spine extruded

Now it's a good idea to give all the bones relevant names because we will have to distinguish them often later on. To rename a bone, select it in **Edit Mode** and press *F12* on your keyboard. We will rename all of the spine bones Spine, Spine.001, Spine.002, and so on, starting with the **Master Bone**.

Now, we'll extrude one more bone for the head:

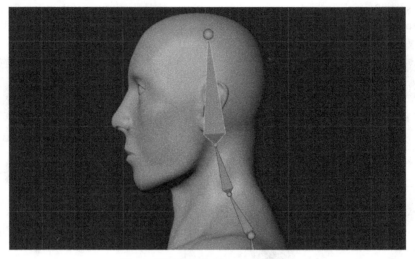

Figure 10.17 – Head bone extruded

Perfect, now let's make the arms and hands.

Arms and hands

Before we start making the actual arms, let's make the clavicle since it helps move the shoulder around and stabilize it.

To make the clavicle, we'll duplicate the bone in the base of the neck using *Shift + D* and position it where we sculpted the clavicles. Remember that the clavicles extend from the base of the neck to the shoulder when positioning and/or scaling the bone to fit your model:

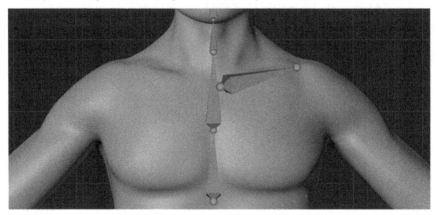

Figure 10.18 – Clavicle bone added

Notice how this bone is disconnected from the others. That is intentional, as we don't need to have all the bones connected and the clavicles are not connected to the spine.

If you notice that your bone has a weird rotation in its long axis, you can adjust the **Roll** parameter in the **Transform** menu, under the **Bone Properties** tab. Make sure to do so in **Edit Mode**:

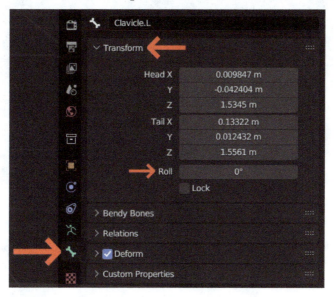

Figure 10.19 – Adjusting the Roll parameter for the bone

While rigging symmetrical things in Blender, we can make just one side and symmetrize it to the opposite side. To do that automatically though, we need to rename the bones we want to symmetrize accordingly. So, at the end of each bone, we'll add the suffix `.L` for the left side or `.R` for the right side. Blender will recognize these suffixes on the names of the bones and automatically symmetrize them to the other side of the mesh. So, in our case, we'll rename the clavicle bone `Clavicle.L` since we're making the left clavicle.

From now on, all of the bones' names will be displayed since it can become confusing to visualize what is being done with a larger number of bones.

Now, due to the complete lack of bones in the ribcage area, Blender will likely struggle when deforming it. That's why we'll add one bone in the chest area, both to help with the deformation of the nearby bones (the clavicle and upper arm) and to allow us to deform the chest to some degree:

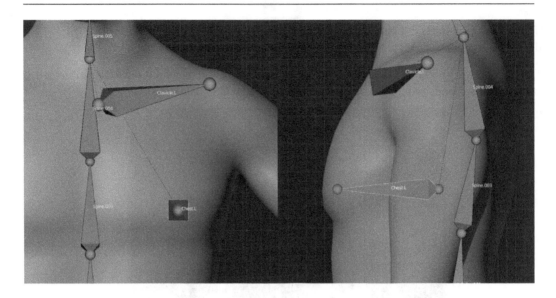

Figure 10.20 – Chest bone added

For the upper arm, we'll duplicate the clavicle bone and position it accordingly, keeping in mind that the humerus is just one bone, so we'll make a bone extending from the shoulders, near the clavicle, to the elbow:

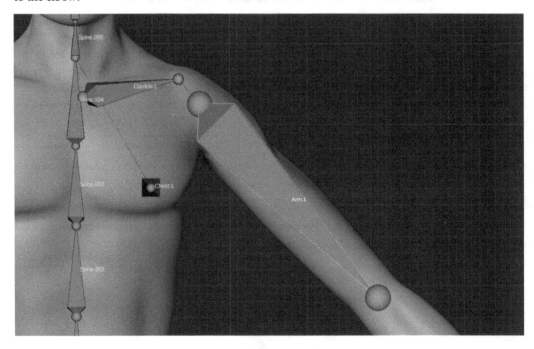

Figure 10.21 – Upper arm bone positioned

Make sure to check the relations between the disconnected bones. To do so, look for dotted lines between them and you'll see them connecting the root of one bone (child) to the tip of the other (parent). The child bone will move with the parent. For example, in this figure, we can see that the **Arm.L** bone is the child of the **Clavicle.L** bone and will move with it. If your upper arm bone doesn't have a dotted line connecting it to the clavicle bone, you'll need to manually parent the bones.

To do this, you can select the child bone, and then, holding *Shift*, select the parent bone and press *Ctrl + P*. Then, select **Keep Offset** from the **Make Parent** menu that will pop up (so as to not make the bones connected). Then, you should see a dotted line appear, connecting the root of the arm bone to the tip of the clavicle bone:

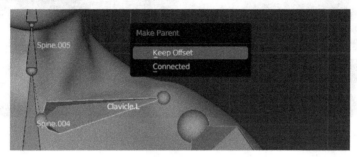

Figure 10.22 – Manually parenting bones

Good, now we can extrude the bone from the upper arm to the forearm, until the wrist. Make sure that all the bones are roughly centered inside the arm as real bones would be:

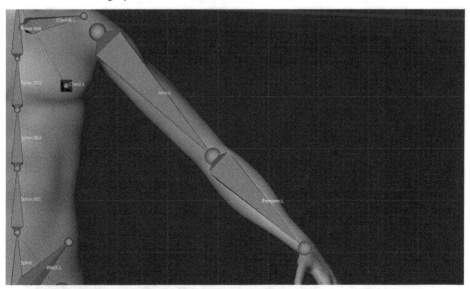

Figure 10.23 – Forearm bone extruded

Now, from the forearm, we'll extrude one bone for the palm, which will control the wrist movement. This bone will extend until the middle of the palm:

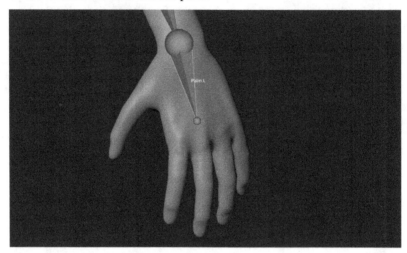

Figure 10.24 – Palm bone extruded

Now, we'll make each finger have a set of four bones (except for the thumb, which will have three) – three for the actual joints on the finger and one for the bones that extend to the palm. We'll duplicate the palm bone for that and remember to keep all of the joints between the bones aligned with the joints on the finger and centered inside each finger:

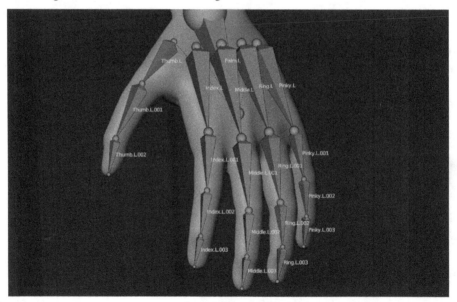

Figure 10.25 – Bones positioned on the fingers

It's important to keep the finger bones disconnected from the palm bone and keep the first bone in each finger's chain as the child of the palm bone.

Perfect – now that we have a full rig for the arms and hands, we can move on to the legs and feet.

Legs and feet

The rigs for the legs and feet are simpler than the rest since the bones on the legs tend to be bigger.

We'll start by extruding the root of the **Master Bone** and positioning it roughly where the pelvis ends. This will allow us to control the rotation of the waist without rotating the entire rig with it:

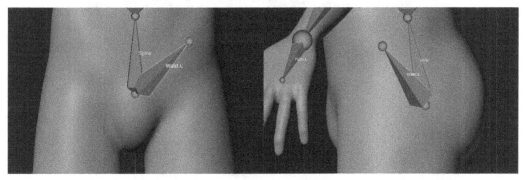

Figure 10.26 – Pelvis bone added

Now, let's duplicate that waist bone and place it in the thigh (disconnected from the pelvis bone), extending from the pelvis to the knee. Make sure to keep the thigh bone centered in the leg and as the child of the master bone:

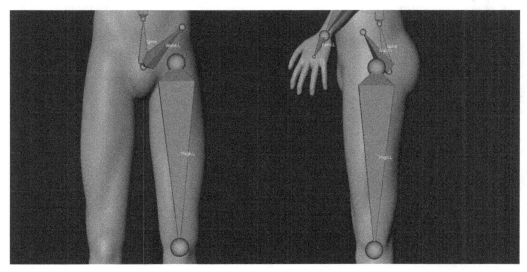

Figure 10.27 – Thigh bone positioned

Good, now we'll extrude the thigh bone to make the shin bone, which will extend to the ankle, as the connection between the lower leg bones and the foot:

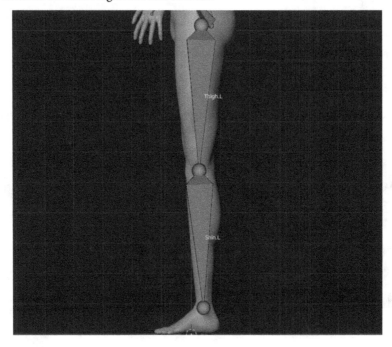

Figure 10.28 – Shin bone extruded

Now we're only left with the foot. We'll extrude the shin bone until near the toes, roughly where the first joint would be, and extrude it one more time for the toes, keeping everything centered in the foot:

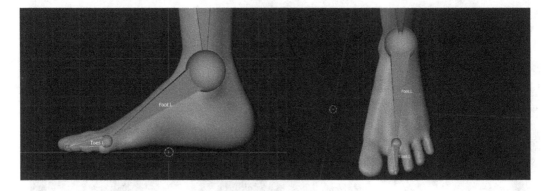

Figure 10.29 – Foot bones extruded

In our case, we didn't put bones into every single toe since they don't move as much and won't be the focus in our case, so we only made one bone to control all the toes but feel free to apply the same logic as we did with the hands in case you need or want control over each individual toe.

Perfect, now we have half of our rig done, and we only need to symmetrize all the bones to the right side. This will only be possible since we named every bone we wanted to be symmetrized with the suffix .L.

To symmetrize the armature, we'll go into **Edit Mode,** select all the bones, *right-click,* and select **Symmetrize** from **Armature Context Menu**, which will pop up:

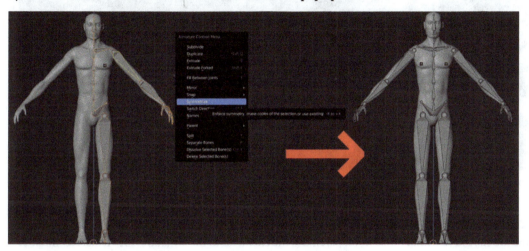

Figure 10.30 – Symmetrizing the armature

Nice, now we have a full rig, but we can't yet move our character. That's because we need to make the character a child of the armature, with an area of influence for each bone. Thankfully, we can let Blender guess what should the area of influence be for each individual bone since it would take too long for us mere mortals to determine exactly how much every single bone should influence. Plus, using math usually gives better results than plain guesswork, so we'll let Blender take the first shot, and correct any mistakes it may have made later.

To parent our character to the armature, we'll select the character, then the armature, press *Ctrl + P,* and select **With Automatic Weights** from the menu that pops up:

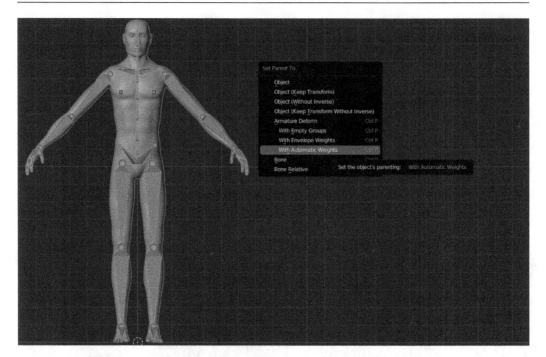

Figure 10.31 – Parenting the character to the armature with automatic weights

Now, we can test the movement of each bone to see whether there are errors in their influence within the mesh. And remember that, upon parenting the character to the armature, the Armature modifier was introduced to our character, and it's important that the modifier is at the right position if there's more than one modifier. Usually having the Armature modifier at the top of the modifier list is what will give the best results.

To test the movements of our character, we need to change to the **Pose** mode from the options in the top left of our Viewport, with the armature selected:

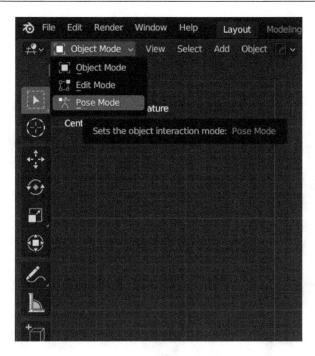

Figure 10.32 – Changing to Pose Mode

Now you can rotate, grab, and/or scale the bones with *R*, *G*, and *S*. Since most bones only rotate, it wouldn't make sense to move around or scale any bones, but this allows for more creative freedom since most of the time we don't have to strictly follow the laws of physics, so sometimes a little bit of movement can help a lot in posing and animation.

While testing, don't take into account whether a bone moves too much just yet; we'll fix that later as right now, every bone has an unlimited range of motion. What you should worry about is whether the bone influences the right area or not. Remember to test every single bone.

In our specific case, the head, clavicle, and upper arm bones have a bigger or smaller area of influence than they should, so we'll fix that by manually adjusting their weights.

Before going into weight painting, though, let's make the bones a little thinner since right now it is a little hard to see how the mesh is deforming from some angles. This step is optional.

To do that, we'll follow these steps:

1. Select the **Armature** in **Object Mode**.
2. Go into the **Object Data Properties** tab.
3. Under the **Viewport Display** menu, select the **Octahedral** option from the **Display As** setting.

4.   Select **Stick** from the menu that pops up.

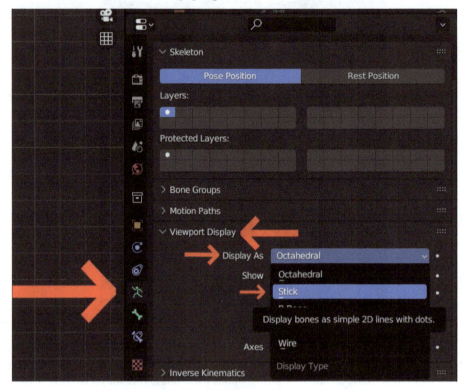

Figure 10.33 – Changing the bones' display to Stick

Good, now that we can see our mesh better, we can go into weight painting.

Weight painting

Weight painting is the process of manually controlling the area of influence (weight) of a specific bone. This method works by applying values that go from 1 (full influence) to 0 (no influence) around the bone. Here's the gradient representing the values:

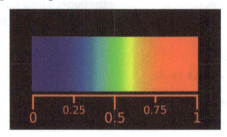

Figure 10.34 – Weight gradient from 0 to 1

To enter **Weight Paint** mode, we have to have our character selected and then switch to **Weight Paint** mode from the menu on the top left:

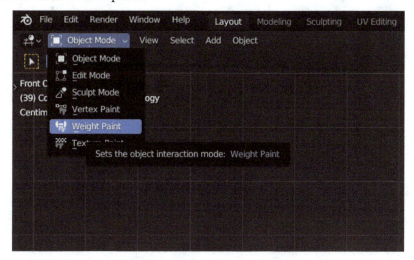

Figure 10.35 – Changing to Weight Paint mode

Upon selecting this mode, most of your character should turn blue (except for the eyes, which are a separate object), and the interface will change to this:

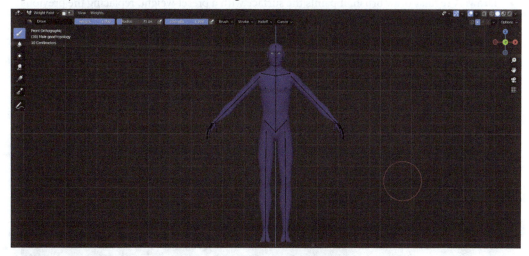

Figure 10.36 – Weight Paint interface

This works very, very similarly to texture painting, so we'll only have a quick look at the main elements in the UI, starting with the column on the left:

Figure 10.37 – Painting modes

These are the brushes that control how the stroke will affect the weights. From the top down, we have the following:

- **Draw**: A regular draw brush. It paints the surface with a preset weight, replacing the previous weight on the surface.

- **Blur**: This makes the transition between two or more nearby weights softer where this brush is applied.

- **Average**: This averages the weight values under the brush's area of influence.

- **Smear**: This smears the weight from one area to another depending on the direction of the stroke. It works similarly to spreading paint that is already on a surface with a brush.

- **Gradient**: This applies a gradient from a set weight down to 0 across the entire surface of the object.

Now let's have a look at the top row of settings:

Figure 10.38 – Top row of Weight Paint settings

Now, the only new parameter we haven't seen yet is the **Weight** slider, which determines the preset weight of some of the brushes we've just seen depend on to apply the weight onto the surface. The rest is mostly identical to texture painting, and you can have a look at brush-specific settings on the **Active Tool** tab:

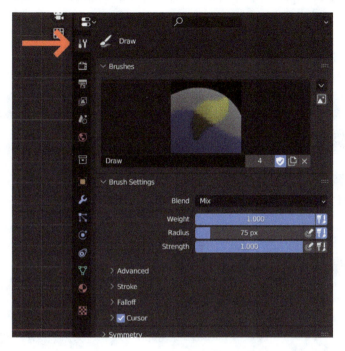

Figure 10.39 – Active Tool tab

For our purposes, though, the specific settings will be left as their defaults, as they will get the job done just fine.

The last UI element worth keeping in mind is **Symmetry**, located at the top right of the Viewport:

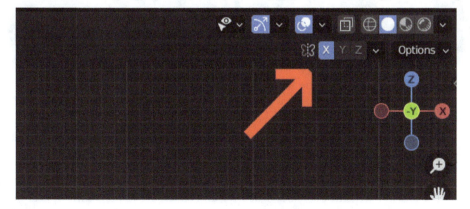

Figure 10.40 – Symmetry option

This automatically applies the same changes made to the weights in a bone with a symmetrical counterpart to that counterpart. It only works on models with symmetrical geometry and with the bones named accordingly.

Alright, so in order to get going with the weight painting, we'll need to know which vertices each bone is affecting and its weight on these vertices. To visualize that, we'll need to go to the **Object Data Properties** tab, and check out the names of the bones under the **Vertex Groups** menu:

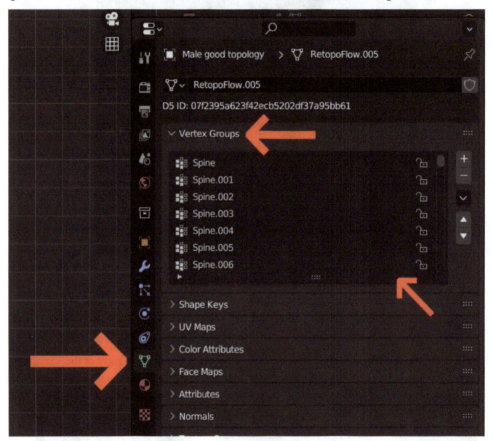

Figure 10.41 – Bone Vertex Groups

Each bone of our armature has a separate **Vertex Group**, which determines which vertices that individual bone will affect when it's moved. The weight of each vertex was also calculated by Blender when we parented our character to the armature with automatic weights.

But wait, we can't see the weight yet. That's because we need to select the bone whose weight we want to see (another reason to rename the bones properly), and its respective weight will appear on the mesh itself:

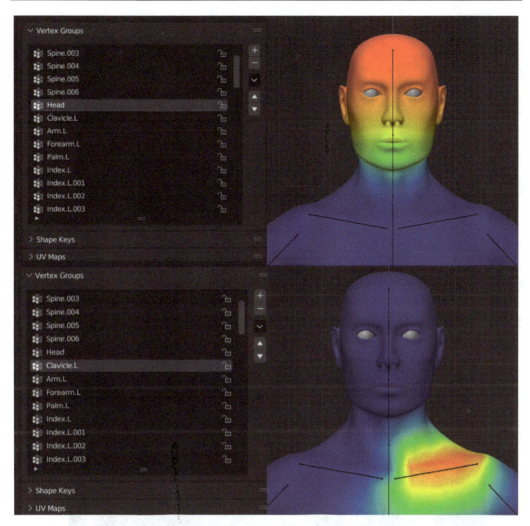

Figure 10.42 – Visualizing the weights of an individual bone

Now we can start painting the correct weights for the parts that Blender didn't get quite right. Starting with the head, we can see in the preceding figure that the head bone is not affecting the whole head, but rather leaving the lower portion of the face and the whole chin with a lower weight, which makes it so that the upper and lower portions of the head don't move together, but rather the lower portion moves less due to its lower weight and distorts the whole face. That is impossible to do without a very, very soft skeleton, and that's not very common among us humans.

For this reason, we'll select the head bone's Vertex Group and paint over the whole head, face, and under the chin using the **Draw** brush at a **Weight** value of 1, so that the head bone affects the whole head uniformly, then using the **Blur** brush to make the transition between the affected and non-affected area smoother:

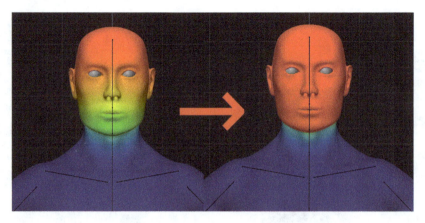

Figure 10.43 – Head bone weight fixed

Remember to check the weight of the nearby bones since overlapping weights can still cause the wrong deformation. In our case, we removed all the weight from the nearest spine bone (**Spine.006**) from the head using the **Draw** brush with a **Weight** value of 0. Keep in mind to test the movement of the bone after each change in the weights to see whether it is deforming correctly.

Now, for the clavicle bone. Let's have a look at how the current weight looks:

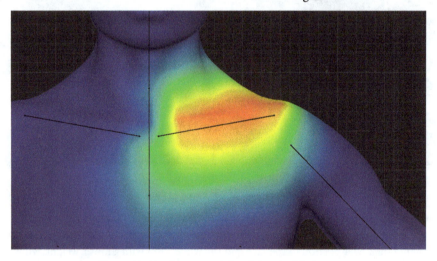

Figure 10.44 – Current weight of the clavicle bone

As you can see, the **Clavicle** bone is affecting a much bigger area than it should, so let's adjust it using the **Draw** brush on the areas further from the actual clavicle and the **Smear** brush on areas closer to it, dragging the weight from the outside to the inside of the **Clavicle** bone's area of influence. Again, check the nearby bones as you test the movements, such as the neck and arm bones:

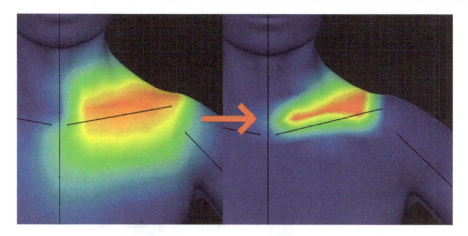

Figure 10.45 – Clavicle weight adjusted

While painting the weight of a bone with a symmetrical counterpart, don't forget to turn on X **Symmetry** before you start painting the weights.

Let's have a look at the difference in the deformation since on the clavicles it might not be as obvious as with the head:

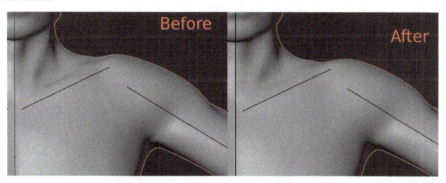

Figure 10.46 – Difference in the deformation of the clavicle bone

In this image, the clavicle was rotated 15° upward, and as you can see, now the shape of the muscles on the neck is way less affected by the movement of the clavicle, and we get a more subtle deformation around the clavicle itself. Looking at references for shoulder movement can also help determine the weight and how that bone should deform its surroundings.

Now the last thing we'll correct is the weights of the upper arm bone, more specifically in the shoulder area. So, let's have a look at how it currently deforms:

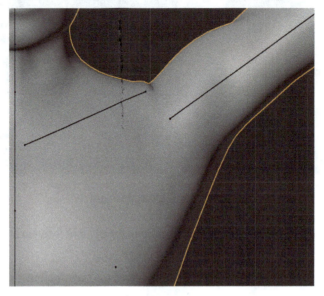

Figure 10.47 – Current shoulder deformation

The arm was raised in combination with the clavicle since this is how it tends to happen in real life. Why it looks wrong is not much of a mystery. The upper arm bone has a bigger area of influence than it should on the front, back, and upper sides, and maybe also on the sides of the body. To be sure, let's have a look at the weight of this bone in the initial position:

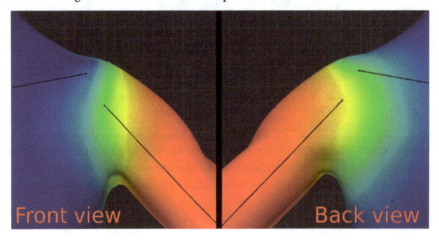

Figure 10.48 – Current weight in the shoulder area

Now we can see the problem a little better. The bone affects the back and sides too much, and the front could use some adjustments too.

We'll shrink the weight from all sides, especially at the back – again, using the **Draw** brush with a **Weight** value of 0 to prevent the areas further from the bone from being influenced at all and using a combination of the **Blur**, **Smear**, and **Average** brushes to make the transition between the shoulder and clavicle bone's influences a bit smoother. Don't forget to turn on the symmetry and test the movement as you go! Let's have a look at a before-and-after of the weight painting for this bone:

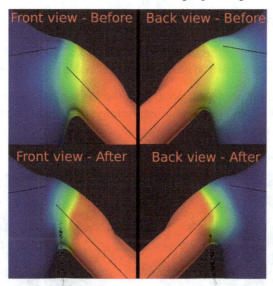

Figure 10.49 – Before-and-after of the weights near the shoulder

That bone now influences the shoulders and the back way less than before and gives a more natural deformation when moved upward. Have a look at the before-and-after of the deformation itself:

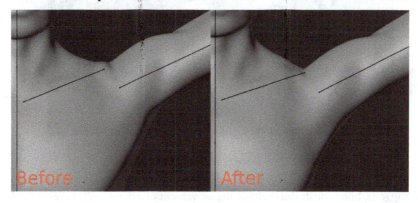

Figure 10.50 – Before-and-after of the deformation of the shoulders

Notice how the deformation is more natural now, with a more precise weight applied to that part of the bone. This is still not perfect deformation, but it's better.

Keep in mind that this rig is extremely simple and therefore has its limitations when it comes to more extreme poses, but it allows for a wide variety of movements, poses, and animations, still being a versatile rig.

Now, with all the wrong weights fixed, we can finally get to limiting the movements of the different joints.

Limiting the movements of the joints

As we explained at the beginning of this chapter, some joints in the human body have limitations that need to be taken into account in order to generate more believable poses and movements.

We'll mostly limit the hinge and pivot joints since they move way less than the rest, and we'll leave some more movement in the other joints to allow for some creative freedom in case we need a movement/pose that is slightly out of reach for a human to do.

To limit a bone's rotation, we'll select it in **Pose Mode**, and go over to the **Bone Constraint Properties** tab:

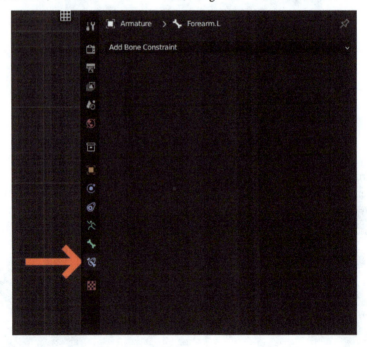

Figure 10.51 – Bone Constraint Properties tab

As the name suggests, bone constraints restrict the movement of a bone, and these can make posing much easier when you know how to use them properly. For now, we'll only use one of them, which is the **Limit Rotation** constraint. We'll select it from the **Add Bone Constraint** menu, at the top of the tab:

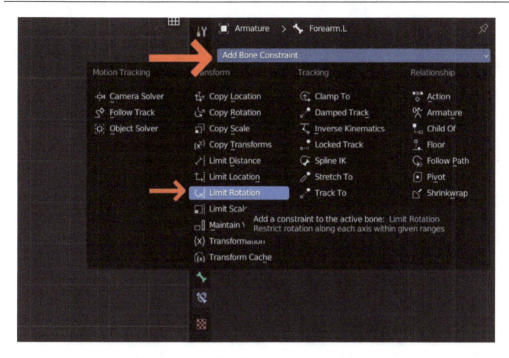

Figure 10.52 – Adding the Limit Rotation constraint

There are very few bones we'll actually limit the rotation of, such as the fingers, knees, and forearms.

Starting with the fingers, in **Pose Mode**, we'll select the first bone of a finger (we chose the index finger), located on the palm, and then add the **Limit Rotation** constraint. This is what the constraint looks like:

Figure 10.53 – Limit Rotation constraint

This is a very simple constraint, and most of the important settings are self-explanatory. When you add the constraint, the bone to which you added it should turn green instead of the normal white. This indicates that the bone has a constraint:

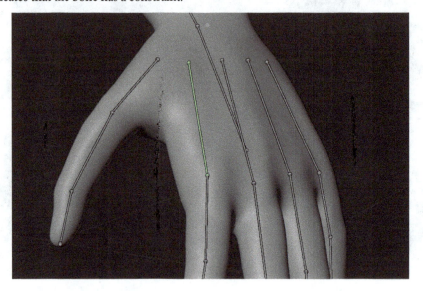

Figure 10.54 – Bone turns green due to the constraint

Now we'll get to the settings. Which axis do we limit the rotation along – *x*, *y*, or *z*? Well, to have a better sense of the orientation of each bone, we can tell Blender to display all of the three axes of every single bone by going into the **Object Data Properties** tab of the armature and toggling the **Axes** option under the **Viewport Display** menu:

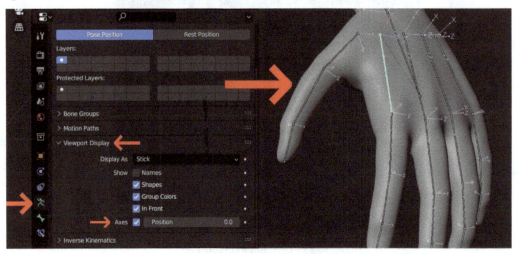

Figure 10.55 – Displaying the axes of each bone

Notice how now all the bones have their respective axes displayed. Keep in mind that, by default, they appear in the root of the bone, not at the tip. This can be adjusted to your liking using the **Position** slider. We'll keep it as the default.

Now we can see better on which axis to limit the rotation of the bones. That first bone of the finger has little movement, and we'll limit its rotation accordingly.

To start off, it doesn't twist, so we'll completely turn off the rotation on the y axis, by turning on the **Limit Y** option for the **Limit Rotation** constraint and leaving the **Min** and **Max** values as 0:

Figure 10.56 – Turning off the Y rotation on the first finger bone

Upon activating this setting, you may immediately notice that that bone will "break" since its rotation will be changed a lot for some reason:

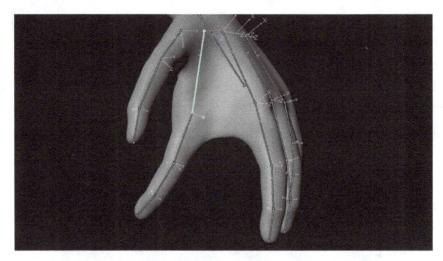

Figure 10.57 – Finger bone rotated by adding the constraint

This happened because Blender is taking into account the **World** coordinates while limiting the rotation, where the z axis points upward. We'll need to change that in the constraint. To do that, simply change the **Owner** setting from **World Space** to **Local Space**, which will make it so that the local coordinates of the bone are taken into account instead of the world ones:

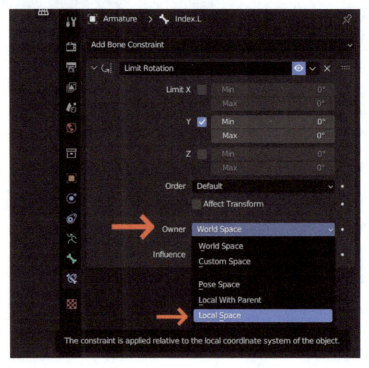

Figure 10.58 – Changing the constraint to Local Space coordinates

Now, that bone should come back to its original rotation and should work correctly, as we'll use each individual bone's coordinates to limit the rotation. Perfect, now that bone doesn't twist anymore.

We can go even further. Knowing that these bones move just slightly in the other directions too. We'll turn on limiting for the x and z axes and adjust the **Min** and **Max** values of each axis according to our needs. We settled with -3° for **Min** and 1.5° for **Max** for the x axis and -3° for **Min** and 3° for **Max** for the z axis.

All the other palm bones for each finger can have the same constraint since they have similar ranges of motion. To duplicate the constraint, we just set up the other bones, select all of the bones that should have that constraint (with the bone we just worked on being the last one selected), then click the little arrow to the right of the constraint's name, and select **Copy to Selected** from the menu that pops up:

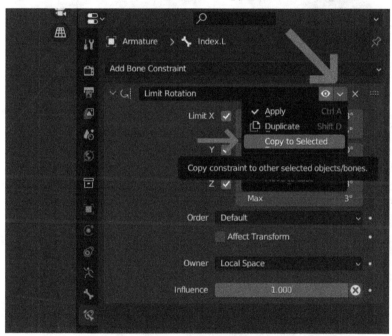

Figure 10.59 – Copying the constraint to the selected bones

Perfect, now the first bones of the fingers have limited rotation. The bone on the pinky finger tends to have a bigger range of motion (around 8° on each side along the z axis, in our case), so feel free to adjust the constraint for the pinky according to your needs (or any finger, for that matter – we're only suggesting values).

For the first joints of the actual fingers, since they're saddle joints, we'll just turn off the twisting and let the other axes move freely, and the constraint looks like this:

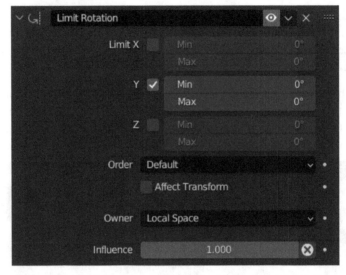

Figure 10.60 – Limit Rotation constraint for the first joint of each finger

This same constraint can also be applied to the first joint of the thumb since it's also a saddle joint. Now, for the second and third joints of all the fingers, we'll *only* allow rotation on the local z or x axis (depending on the **Roll** setting of the bone), whichever only allows it to rotate up and down since they are hinge joints. Their constraint looks like this:

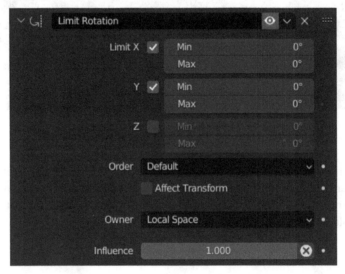

Figure 10.61 – Constraint for the last two joints of the fingers

With this, all the fingers should have appropriate movement. Don't forget to copy these constraints to the other hand!

Next, we'll limit the rotation of the forearm, in a similar way to what we did for the fingers. Let's check the orientation of the forearm bone to work out for which axis we'll limit the rotation:

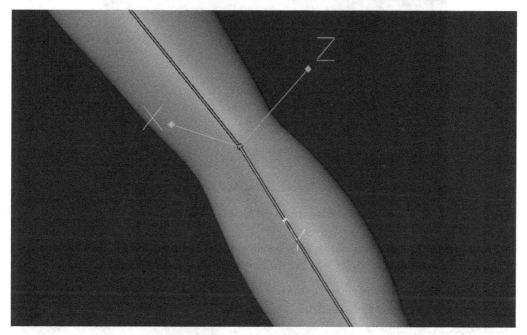

Figure 10.62 – Forearm bone orientation

In this case, we'll only turn off the rotation for the local x axis since it bends along the z axis and we still want some twisting. Here's how the forearm constraint looks:

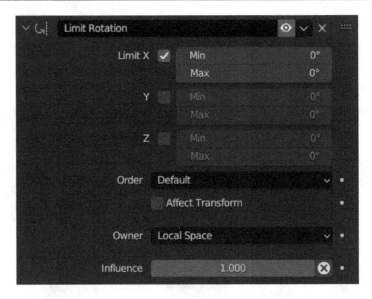

Figure 10.63 – The forearm bone's Limit Rotation constraint

Now we can't break the forearm accidentally by bending it upward or downward, since the limit for both sides is 0°. Don't forget to apply this same constraint to the other forearm as well.

The last bone we'll add constraints to is the shin bone. This will also have a pretty similar setup to the fingers. First, let's check the orientation of the shin bone:

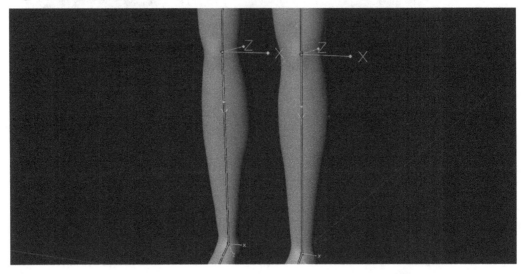

Figure 10.64 – Shin bone's orientation

The shin bone doesn't twist or bend sideways, so according to this figure, we'll have to turn off the rotation on the local *z* and *y* axes. Here's what the constraint looks like:

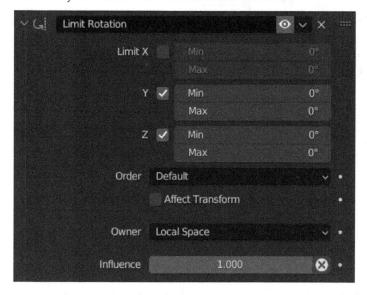

Figure 10.65 – The shin bone's Limit Rotation constraint

Perfect, now we have finished limiting the rotation of the necessary bones and our rig is ready. Feel free to go back and explore with more constraints and/or tweak the constraints we have already added.

We suggest posing the character yourself in a few different poses to get the hang of how posing the finished rig works. Use references as you do so since remembering how all the different parts of the body move together is very difficult.

Now, as soon as you start actually posing your character, you'll notice it is an extremely tedious and annoying process since you have to move every single bone individually, which probably got you wondering how a sane person can animate entire characters like this without going to a therapist. Well, the answer is very simple. The rig we just made uses a method called **FK** to move.

FK is a system that moves the bones using parenting only, so if we rotate a parent bone, all the children bones rotate with it. For example, let's say that **Bone.001** is the parent of **Bone.002**, which is the parent of **Bone.003**. If we rotate **Bone.001**, all the others will rotate along with it. If we rotate **Bone.002**, **Bone.003** will rotate along with it while **Bone.001** remains still, and if we rotate **Bone.003**, none of the other bones will rotate with it since it's not the parent of any bone. In a way, we can say that the movement on an FK rig goes from parent to child. This means that we have to move every individual bone in order to pose a character. This is how every rig starts out.

Most of the time, however, when we need more complex posing or animation, we tend to use FK's fancy cousin: **IK**.

IK works by having a controller bone at the end of the chain and using that to move and rotate all of the bones in the chain at once, not necessarily following the FK's parent-to-child logic for the movement, but rather being almost the opposite since we control the entire chain of bones using a controller at the end. This makes it much easier to pose the hands and feet, for example.

For that reason, let's add some IK controls to our current rig.

Adding IK to our rig

In our case, we'll add IK to the arms and legs only, and since we will control the movement using controller bones, we'll actually remove the constraint we previously added to them with the FK rig.

Let's start with the legs.

Legs

For the legs, we'll need a couple of extra bones extruded, one at the ankle and another at the knee. So go into **Edit Mode** on the **Armature** and extrude them (do this for both legs so that they have separate controls):

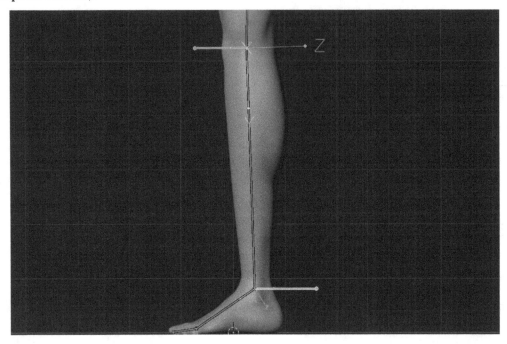

Figure 10.66 – Bones extruded at the ankle and knee

It's important to extrude the heel to the back and the knee forward since this helps both with visualization and functionality, which we'll explain later. These bones don't have any influence on the deformation of the character since they don't have any weight assigned to them. That's what we want since we'll use the controller on the heel as the source of movement for the entire leg and the knee bone to control where the knee of our character points. Make sure to name these new bones correctly.

Then, with the knee and heel controllers selected, in **Edit Mode**, we'll remove their parenting with any bones in the rig by pressing *Alt + P*, then selecting **Clear Parent** from the menu that pops up; this will make it possible to move these bones independently:

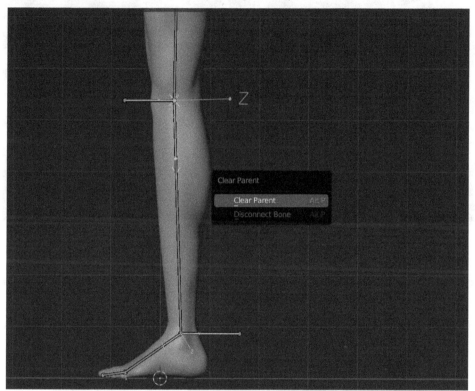

Figure 10.67 – Clearing the parents from the controller bones

Now, move the knee controller away from the knee since this allows for more precision when controlling the rotation of the knees:

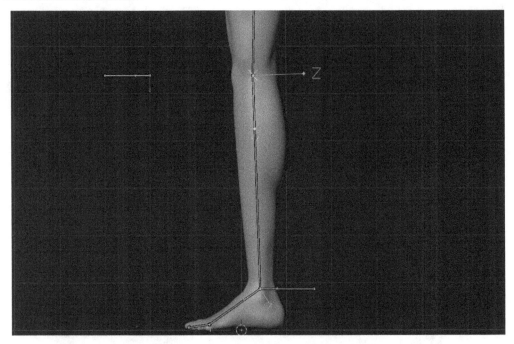

Figure 10.68 – Knee controller moved away from the knee

Now that we have all the controllers, let's add IK to the beginning of the chain of bones we want to be affected by IK, which in our case is the shin bone.

Now, in **Pose Mode**, we'll select the shin bone and add the **Inverse Kinematics** constraint. Yes, it is a constraint since it influences the movements of the bones.

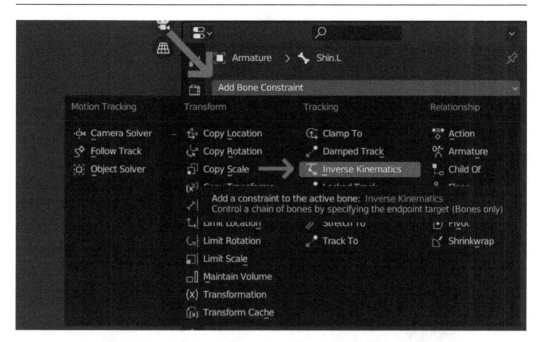

Figure 10.69 – Adding the IK constraint to the shin bone

Now, upon adding the constraint, you'll be presented with this on the **Bone Constraints** tab:

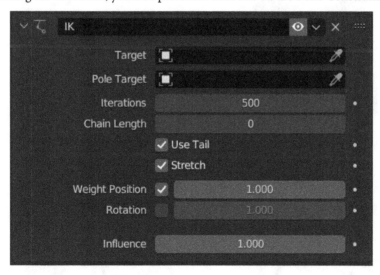

Figure 10.70 – IK constraint

Upon adding this, the bone you added it to should turn orange. That means that the constraint is applied but not actively influencing the bones.

The first impression might not seem too friendly, but don't be scared, we won't change most of the settings. For a target, we'll select our **Armature**, then another option will appear underneath it, asking for a bone:

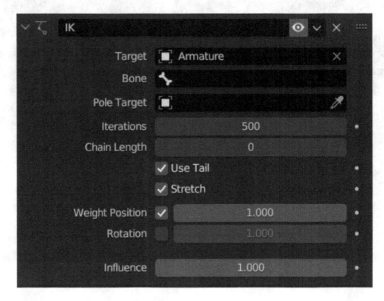

Figure 10.71 – Choosing the Armature as the Target for the IK constraint

Under **Bone**, we'll choose the control bone we added to the heel:

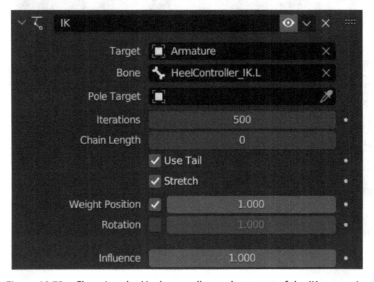

Figure 10.72 – Choosing the Heel controller as the target of the IK constraint

Now that we have set our control bone, the constraint is affecting the bones, but not in the way we want it to. Right now, the whole rig moves chaotically when we move the controller.

To fix this, we need to set a **Pole Target**, which will give the IK constraint somewhere to point to, so to speak. For the **Pole Target**, we'll also set the **Armature** as the object since the knee control is part of the armature as well, and just like before, the **Bone** option will appear underneath it. We'll select the knee controller as the **Pole Target**:

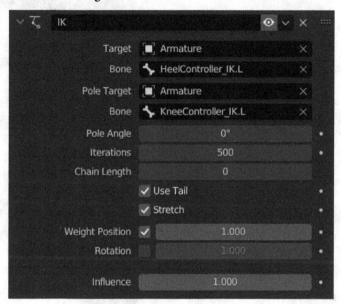

Figure 10.73 – Setting Pole Target

Unfortunately, it's not fixed yet, as the IK is still affecting the whole armature since we didn't tell Blender how many bones to affect. To do that, we'll change **Chain Length** to the number of bones we want the IK constraint to affect. At 0, it affects the whole armature. At 1, the IK will only affect the shin bone since it is the one that has the actual constraint applied to it. At 2, it will affect the shin and the thigh bones since the thigh bone is connected to the shin bone. We want to affect the leg only, so we'll keep **Chain Length** at 2 bones.

Now you'll notice that the leg looks broken, again:

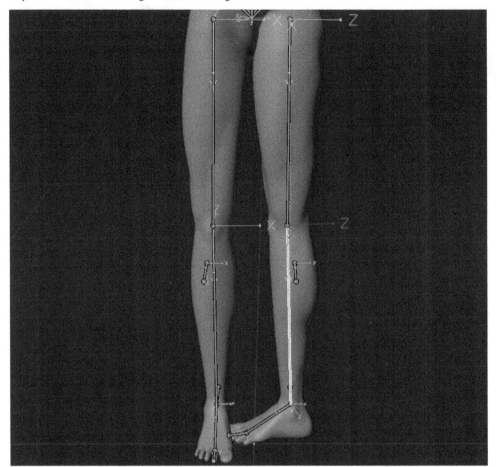

Figure 10.74 – Twisted leg after adjusting Chain Length

As you can see, the whole leg was twisted by 90° from its correct rotation. That happens because we still have to tell Blender which part of the bone to point at the knee control. To do that, we'll have to adjust the **Pole Angle** setting. Usually, an angle of 90° or -90° does the trick well, but sometimes, 180° or -180° is required. For this leg, -90° worked. Now, the leg should be pointing forward, and in **Pose Mode**, when we move around our controller bones on the heel using *G*, we should be met with a very pleasant surprise: a leg that can be entirely controlled by moving two bones around:

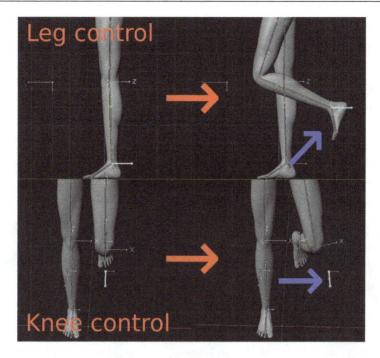

Figure 10.75 – Full leg IK movements

As you can see, all of the leg bones move in a natural way, making it much, much easier to configure poses such as crouching, for example. Here are the final settings for the IK constraint on this leg:

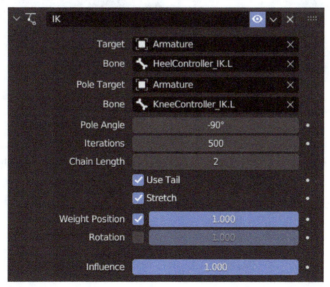

Figure 10.76 – Final settings for the IK constraint on the left leg

Now, we still need to follow this same process for the other leg, selecting the equivalent bones for the constraint, which makes simply transferring the constraint from one leg to another impossible, not to say incredibly annoying.

Luckily, Blender offers an easier way of adding IK to a chain of bones. For the other leg, in **Pose Mode**, we'll do the following:

1. Select the control bone in the heel.

2. Select the shin bone (or whatever bone to which you'd add the IK constraint).

3. Press the *Shift + I* shortcut.

4. Select **To Active Bone** from the menu that pops up.

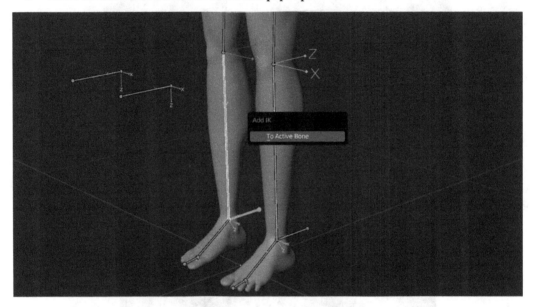

Figure 10.77 – Automatically adding the IK constraint to the shin bone

This will automatically add the IK constraint to the shin bone and set up the control bone as the target, which only leaves us with the task of changing **Chain Length** to 2 and setting **Pole Target** as the knee control.

Perfect, now that the legs are working, let's move on to the arms.

Arms

The process of adding IK to the arms is extremely similar to the legs, only diverging from it on the extruded bones.

We'll extrude one more bone at the wrist and one at the elbow, both being extruded to the back, and clear any parenting relationships with any bones in the armature, just like we did with the knee and heel controllers:

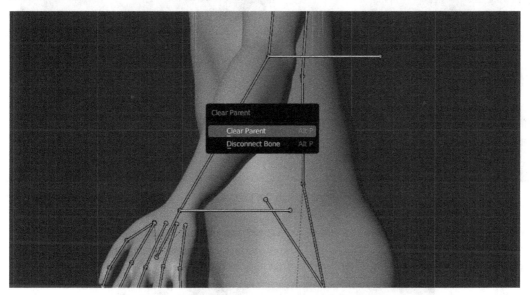

Figure 10.78 – Extruding the controller bones to the arms and clearing their parenting

Now, as we did with the knee controller, we'll move the elbow controller further back, away from the elbow. This bone will be our **Pole Target** when we add the IK constraint. Remember to remove any constraints from the arm bones you'll control with the IK constraint.

Now, in **Pose Mode**, we'll add the IK to the arm by doing the following:

1. Select the wrist controller, which will control the movement of the entire arm.
2. Select the **Forearm** bone, which is the beginning of the chain.
3. Press *Shift + I*.

4.   Select **To Active Bone** from the menu that pops up.

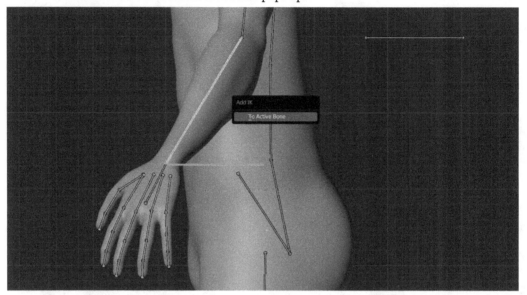

Figure 10.79 – Adding IK to the forearm automatically

With the IK constraint automatically added, we just need to set **Pole Target** as the elbow controller, change **Chain Length** to 2 since we just want to control the forearm and upper arm bones, and adjust **Pole Angle**, in our case, to -180°. Here are the final settings for the IK constraint on the arm:

Figure 10.80 – Final settings for the forearm IK constraint

Don't forget to apply the same constraint to the other arm.

Perfect, now we have fully functional arms that can be controlled using only two bones, just like the legs:

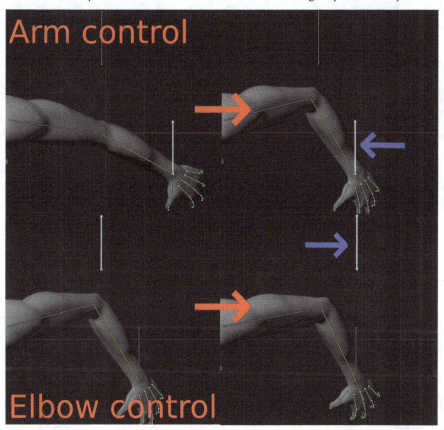

Figure 10.81 – Full arm IK movements

As you can see, when we move the wrist control, we move the entire arm, and when we move the elbow control, we control the rotation of the elbow.

We're mostly done with our rig now; however, if you plan on moving the entire character around, using the **Master Bone** at the waist won't move the entire rig anymore because of the IK controllers we added. We will need to add another master bone. To do that, in **Edit Mode**, we'll press *Shift + A*, which will add another bone at the **3D Cursor**. We'll position that bone right between the feet of our character:

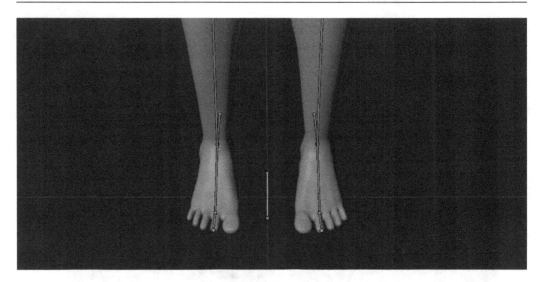

Figure 10.82 – New master bone added and positioned

Now, we'll parent all the IK controllers and the current **master bone** to the new master bone by selecting all of the bones we'll parent, then the new master bone, pressing *Ctrl + P*, and selecting **Keep Offset** from the menu that pops up:

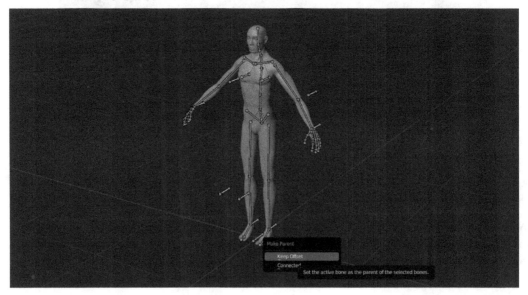

Figure 10.83 – Parenting the IK controllers and previous master bone to the new master bone

With this, we can move our entire character at once using just one bone, which makes it easier to animate characters walking, flying, crawling, or anything that requires the entire rig to change its location in 3D space.

Perfect, now we have a fully movable character with limbs that are a lot easier to pose!

Remember, though, that this is a simple rig, and, even with IK, it has some limitations. We won't go over more complex rigging since it gets very complex extremely quickly. This rig is enough to get you started with rigging many characters, as it's simple to set up and understand, as well as versatile, offering a wide range of possible poses. Again, feel free to explore the functionality of the rig, adding constraints to bones, adding extra support bones… We encourage you to explore.

However, sometimes, we just need more extreme movements that our current rig doesn't allow as it is right now, and that's where Blender comes to our rescue, with the free in-built Rigify add-on!

Generating a full rig with the Rigify add-on

Rigify is a built-in, free add-on that has a library of pre-made rigs, ranging from human to animals such as wolves, birds, fish, and sharks. These rigs are equipped with IK, controls, and deformation bones to assist us while posing. We can tweak those rigs to fit our needs and make it possible to pose our characters and creatures with much more ease than making a custom rig from scratch.

To enable this add-on, we will do the following:

1. Select the **Edit** menu from the top left on our viewport.
2. Select **Preferences**.
3. Go into the **Add-ons** tab.
4. Search for Rigify in the search bar in the top right.
5. Enable the **Rigify** add-on by checking the box to the left of its name.

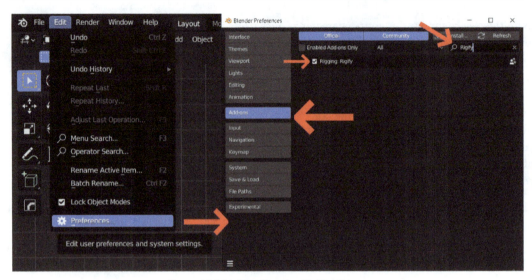

Figure 10.84 – Enabling the Rigify add-on

Now, with the Rigify add-on enabled, when we press *Shift + A* to add an armature, we'll see extra options in the **Armature** submenu:

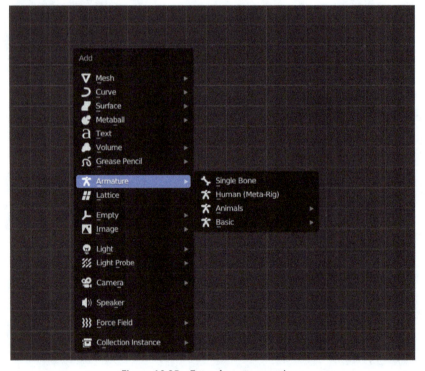

Figure 10.85 – Extra Armature options

These are the extra rigs that Rigify has. We'll add the **Human (Meta-Rig)** rig, so select it. This is what the rig looks like:

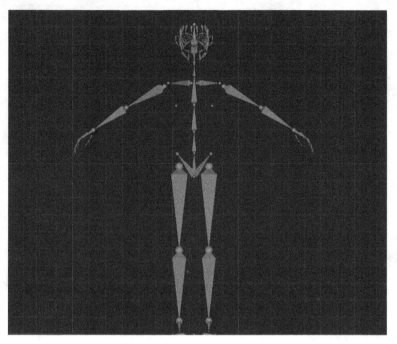

Figure 10.86 – Rigify's humanoid rig

You'll notice how this rig looks very similar to the one we made from scratch, but the real difference comes in the functionality, when we actually pose the character. Also, notice how this rig has a facial rig that we can use. Feel free to use it in your character, but in our case, we'll delete all the facial bones in **Edit Mode**, as it won't make much difference for our purposes. The logic to make it work is the same, though, so don't worry.

Now with all the facial bones deleted, we will move the armature to the same position as our character and start positioning every bone in correspondence with its respective joints in the human body. Don't forget to turn on the **In Front** option from the **Object Data Properties** tab before positioning the bones and enable **X Symmetry**. Since we already made an entire rig from scratch, positioning every bone correctly should be fairly easy. Here's how it should look:

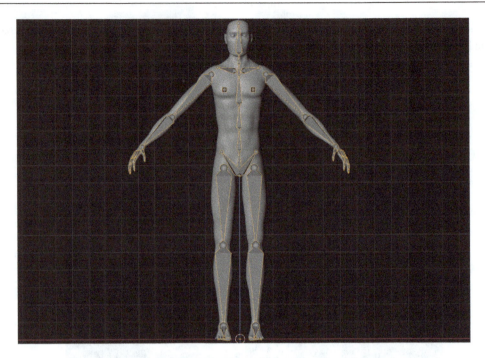

Figure 10.87 – Rigify's humanoid rig positioned in the body

Don't forget to keep the roots and tips connected since some bones in the rig may appear to be connected but are not, so when we move them, we disconnect the bone. When that happens, we have to select both the tip of one bone and the root of the other to move that joint correctly. Here's a demonstration:

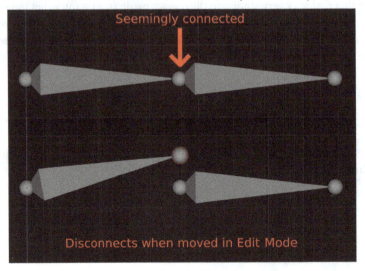

Figure 10.88 – Seemingly connected bones

These types of connections are usually on the hands in Rigify's human rig, so we need to keep an eye on these joints and make sure to keep them connected by selecting both the tip of one bone and the root of the next. Otherwise, the rig might not work as intended.

Perfect, now with the bones correctly positioned, we can do the following:

1. Select **Armature** in **Object Mode**.

2. Go into the **Object Data Properties** tab.

3. Scroll down until we see the **Rigify** menu.

4. With the **Armature** still selected, select **Generate Rig**.

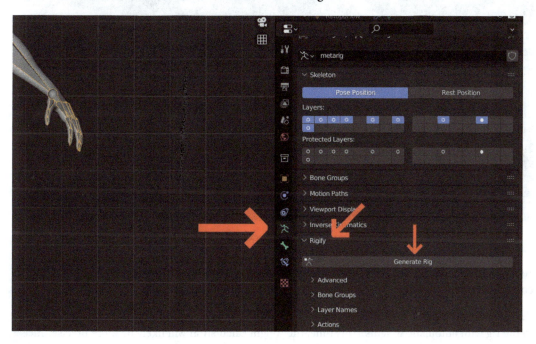

Figure 10.89 – Generating a rig using Rigify

As soon as you click it, a new rig will appear with bones that have different shapes from Blender's default shapes:

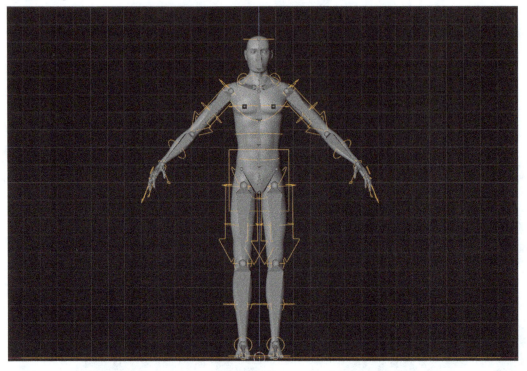

Figure 10.90 – Rigify's generated rig

As you can see, the rig fits perfectly with our character. That's because we positioned each bone correctly inside the mesh.

Perfect, now we can delete the original rig we used to adjust the joints and parent the character mesh to the rig with automatic weights. This time, the weight calculation will likely be more precise since Rigify adds extra bones when we generate the rig to support the deformation of the bones. This means we're less likely to have to do any weight painting after we parent the character with the rig.

In addition to the full rig, we have both IK and FK controls to help us have more control over the mesh, as well as **Tweak** controls in case we want to deform the limbs. We can differentiate the different types of controls by their color when in **Pose Mode**:

- *Red, yellow, and orange*: IK controls
- *Green*: FK control
- *Blue*: Tweak controls
- *Purple*: Master bone (or root bone)

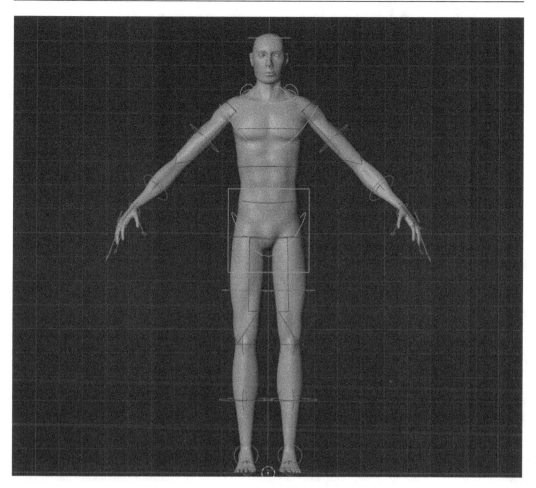

Figure 10.91 – Types of controls in Rigify's human rig

Most of these bones are controlled with rotation, but some can also be controlled by changing their location in space. The fingers can be controlled by scaling the long orange controllers using the *S* shortcut. Sometimes, the green controllers on the fingers can assist while posing.

If you think the rig is a bit cluttered, we can hide some of the layers it has by pressing *N* and navigating to the **Item** menu on the top right, where we'll find all the layers of the generated rig:

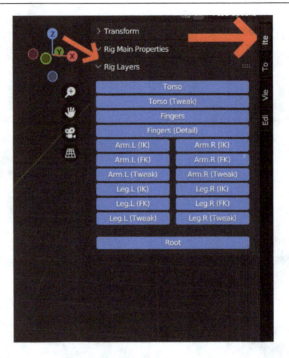

Figure 10.92 – Rig layers

We can disable any layer we want/need by deselecting it. All the layers are conveniently named for ease of use so that we know exactly what we're hiding.

Another aspect to take into account is how to switch from IK controls (which are on by default) to FK controls (which are off by default). To do that, in **Pose Mode**, we'll select the FK control we want to switch to, and again in the **Item** menu, we'll navigate to the **Rig Main Properties** submenu, where we'll slide the **IK-FK** slider for the specific FK bone group we selected:

Figure 10.93 – IK-FK switch

There are more properties to Rigify's rigs, but they aren't as relevant to us. Like always, we encourage you to explore.

Perfect, now we have full control over our character with a rig that allows us to set up more complex poses and animations with more ease than the rig we built from scratch.

Congratulations! Now this finished and posable character is suitable for a variety of use cases such as animations, still renders, and for use as a base for making other characters… Have fun with it!

Summary

In this chapter, we covered in depth how to rig a humanoid character, where to place every bone, and how to make the rig deform the mesh. We also covered how to correct improper deformation using weight painting, as well as how to limit the rotation of certain bones to respect the range of motion in the human body.

Additionally, we covered how to add IK to our rig in order to pose our character with more ease instead of having to rotate and/or move every bone individually. At the end, we covered how to generate and properly adapt a fully functional rig using the free add-on Rigify, which is built into Blender.

Now, you have the technical knowledge to make a wide variety of assets for multiple purposes. What's left is to work on your skill sets, practice, and improve as an artist.

So, what now?

11

Further Development as a 3D Artist

At this point, you have the technical knowledge needed to make a wide variety of assets for a multitude of purposes, and what's left is to improve your skills, initially focusing on improving what you like the most about the creative process in order to get the most out of it.

In this chapter, we'll cover the following:

- Tips on how to quickly improve as a 3D artist
- The importance of feedback

It's important to note that this book only scratched the surface of what is possible with Blender, giving you a solid base on which to start building your skills. But how?

Improving quickly

Regardless of which part of the creative process you enjoy the most, improving your skills will likely be a long and gradual process, and more often than not, we feel discouraged from trying, especially when we see all of the stunning art we're usually presented with on social media. But even though it's a long process, that doesn't mean there's no way of accelerating it a bit. So, here are a few suggestions on how to build up your skills quickly, starting with the biggest and most accessible: tutorials.

Free online tutorials

Chances are, if you've ever looked for 3D-related content on the internet, you've come across Blender content. Now is the golden era for Blender, with thousands and thousands of content creators sharing their knowledge of the program for free on social media. If you're having any technical problems or bugs, or you simply want to look for answers on how to do something you don't know yet, you'll definitely find what you're looking for somewhere.

It's also important to surround yourself with content related to your specific niche. Whether you like sculpting characters, sci-fi hard surface modeling, environment creation, product modeling, or interior design, look for content that is relevant to you. It's OK to feel a bit lost in the beginning, but eventually, everyone finds their preferred niche. So, if you have any doubts, don't feel discouraged from looking it up because the chances of finding something are extremely high.

However, while free tutorials are a great source of quick information, they can only get you so far. If your intention is to get serious and more in depth about something, it's a good idea to invest some time in dedicated courses.

Dedicated courses

Many of the same creators making free tutorials offer courses in which they share their knowledge on a specific topic they mastered to a deeper extent, and while many of these courses are paid (and depending on the level of expertise of the creator and how in depth they go, they can be quite costly), they usually have the potential to give you much more knowledge than a tutorial series, due to the higher focus on one specific topic or aspect of the creative process.

But regardless of which areas you choose to study, you should be looking for feedback.

The importance of feedback as an artist

Feedback is one of the most important aspects of improving what you create, especially as a beginner, when you don't have enough skill to notice some of the mistakes you make. Even professionals look for feedback, albeit with less frequency. When we work on something for several hours, we tend to become blind to our own mistakes.

The perfect (but not the only) time to ask for other people's opinions is when we're unsure of whether we've made a mistake. That can help us to avoid headaches later on when it becomes harder or too late to fix things. But to whom do we go for feedback?

It's important to look for feedback from people who have at least some knowledge of 3D modeling. There are entire communities dedicated to it across the internet/social media, and those are the places where you'll get the best feedback. Most people are willing to share a bit of their knowledge to help someone improve their skills. The Blender community is big and rapidly growing, meaning it's the best place to go if you encounter a technical problem specific to Blender. There is absolutely no shame in reaching out for help, since no one can learn everything alone.

However, even in dedicated communities, there will be good feedback, which you should take into account, and bad feedback, which you shouldn't, and sometimes it can be difficult to differentiate between the two.

Differentiating between good and bad feedback

It's not a surprise that we'll need feedback throughout our entire journey, but which feedback should you listen to?

Good feedback

Good feedback is feedback designed to help you grow: what we're looking for when it comes to feedback is constructive criticism of our art.

Usually, this type of feedback comes with an explanation of why, for example, something in your model looks off and how to fix it, rather than simply pointing out that something looks wrong, which most people can do. When people are willing to elaborate on their opinions of our work, the advice likely has good intentions, and while some critiques might not be valid in their entirety due to our specific situation, more often than not we can extract something useful.

But since we're talking about the internet after all...

Bad feedback

We'll receive tons of feedback along the way, and if your feedback is coming directly from the internet, which it likely will when it comes to 3D modeling, there will likely be bad feedback, which you shouldn't listen to, hiding among the good feedback.

Bad feedback is likely to slow your progress, and while some bad feedback is meant to tear you down and so is pretty obvious to spot, such as when it contains insults or downright lies, some bad feedback is harder to spot.

Typically, this type of feedback comes in the form of non-elaborated opinions, such as pointing out that your piece looks weird and not even explaining why, let alone suggesting a way to improve it. This can leave someone clueless about what is supposedly wrong with the project, while at the same time keeping that thought in the back of their mind, which might lead them to change something unnecessarily. It's extremely important to filter out the bad feedback in order to keep your progress steady. This can seem obvious, but sometimes we can forget the obvious.

Of all the advice given in this chapter, perhaps the most important piece is to never stop practicing, as practice makes perfect, and persistence is key to success. Never stop learning new things and don't be afraid of exploring new ground.

Summary

In this chapter, we covered some important pieces of advice to help you start building your skills beyond what was covered in this book. We covered free and quick sources of knowledge as well as longer and more specialized ones, which typically expand your expertise more. We also covered the importance of feedback and how to differentiate between good and bad feedback, especially when it comes from people on the internet.

Now it's time to practice your skills and have a go at all of the goals that inspired you to read this book. Good luck on your journey.

Index

Packtpub.com

Subscribe to our online digital library for full access to over 7,000 books and videos, as well as industry leading tools to help you plan your personal development and advance your career. For more information, please visit our website.

Why subscribe?

- Spend less time learning and more time coding with practical eBooks and Videos from over 4,000 industry professionals

- Improve your learning with Skill Plans built especially for you

- Get a free eBook or video every month

- Fully searchable for easy access to vital information

- Copy and paste, print, and bookmark content

Did you know that Packt offers eBook versions of every book published, with PDF and ePub files available? You can upgrade to the eBook version at packtpub.com and as a print book customer, you are entitled to a discount on the eBook copy. Get in touch with us at customercare@packtpub.com for more details.

At www.packtpub.com, you can also read a collection of free technical articles, sign up for a range of free newsletters, and receive exclusive discounts and offers on Packt books and eBooks.

Other Books You May Enjoy

If you enjoyed this book, you may be interested in these other books by Packt:

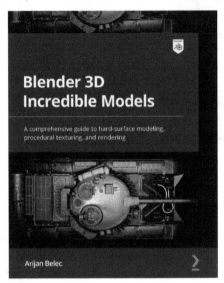

Blender 3D Incredible Models

Arijan Belec

ISBN: 9781801817813

- Dive into the fundamental theory behind hard-surface modeling.
- Explore Blender's extensive modeling tools and features.
- Use references to produce sophisticated and accurate models.
- Create models with realistic textures and materials.
- Set up lighting and render your scenes with style.
- Master the use of polygons to make game-optimized models.
- Develop impressive animations by exploring the world of rigging.

Blender 3D By Example

Xury Greer | Oscar Baechler

ISBN: 9781789612561

- Explore core 3D modeling tools in Blender such as extrude, bevel, and loop cut
- Understand Blender's Outliner hierarchy, collections, and modifiers
- Find solutions to common problems in modeling 3D characters and designs
- Implement lighting and probes to liven up an architectural scene using EEVEE
- Produce a final rendered image complete with lighting and post-processing effects.
- Learn how to use Blender's built-in texture painting tools.

Packt is searching for authors like you

If you're interested in becoming an author for Packt, please visit authors.packtpub.com and apply today. We have worked with thousands of developers and tech professionals, just like you, to help them share their insight with the global tech community. You can make a general application, apply for a specific hot topic that we are recruiting an author for, or submit your own idea.

Hi!

I am Vinicius Machado Venancio, author of *Blender 3D Asset Creation for the Metaverse*, I really hope you enjoyed reading this book and found it useful for kickstarting your journey of 3D modeling in Blender.

It would really help us (and other potential readers!) if you could leave a review on Amazon sharing your thoughts on this book.

Go to the link below or scan the QR code to leave your review:

`https://packt.link/r/1801814325`

Your review will help me to understand what's worked well in this book, and what could be improved upon for future editions, so it really is appreciated.

Best Wishes,

Download a free PDF copy of this book

Thanks for purchasing this book!

Do you like to read on the go but are unable to carry your print books everywhere? Is your eBook purchase not compatible with the device of your choice?

Don't worry, now with every Packt book you get a DRM-free PDF version of that book at no cost.

Read anywhere, any place, on any device. Search, copy, and paste code from your favorite technical books directly into your application.

The perks don't stop there, you can get exclusive access to discounts, newsletters, and great free content in your inbox daily

Follow these simple steps to get the benefits:

1. Scan the QR code or visit the link below

https://packt.link/free-ebook/9781801814324

2. Submit your proof of purchase

3. That's it! We'll send your free PDF and other benefits to your email directly

www.ingramcontent.com/pod-product-compliance
Lightning Source LLC
Chambersburg PA
CBHW060920060326
40690CB00041B/2810